The Croad Langshan
Beauty Based Upon Utility

by Sardius Hancock

with an introduction by Jackson Chambers

Self Reliance Books

Get more historic titles on animal and stock breeding, gardening and old fashioned skills by visiting us at:

http://selfreliancebooks.blogspot.com/

Introduction

I am pleased to present yet another title on Poultry.

The work is in the Public Domain and is re-printed here in accordance with Federal Laws.

As with all reprinted books of this age that are intended to perfectly reproduce the original edition, considerable pains and effort had to be undertaken to correct fading and sometimes outright damage to existing proofs of this title. At times, this task is quite monumental, requiring an almost total "rebuilding" of some pages from digital proofs of multiple copies. Despite this, imperfections still sometimes exist in the final proof and may detract from the visual appearance of the text.

I hope you enjoy reading this book as much as I enjoyed making it available to readers again.

Jackson Chambers

FIRST AND SPECIAL, MADRESFIELD.

Bred by Mr. T. RICHARDS.

CONTENTS

ILLUSTRATIONS

The Editor's Foreword.

MR. SARDIUS HANCOCK.

BY the issue of this Textbook an effort has been made to secure for the Croad Langshan its true place under the new conditions. A glance at its contents will show that not only has the exhibition side received the careful and accurate treatment its importance demands, but that the utility side has been developed with a systematic attention to detail that might have seemed singular in pre-war days. In the articles falling under this head there is a designedly free discussion of the standard. This is as it should be. We Britons live by debate. The Standard of Excellence is the law for breeders and judges alike, but it is well to recognise the essential difference between a breed and its standard ; and that the standard is for the breed and not the breed for the standard. No charge is more frequent nor less justified than that a deviation from the standard is a mark of impurity, for the experience of generations of breeders goes to prove that while an impurity will manifest itself in a deviation, a deviation may be procured by simple selection. Now, while the Langshan is un-doubtedly a pure breed, a study of Miss Croad's book reveals the fact that these original Langshans were not precisely alike ; and it is highly probable that the differences found to-day among Croad Langshans are not so great as among the original importations of half a century ago.

We must remember that the constant effort of fanciers is to breed likeness. If our efforts were directed to breed difference, two or more varieties could, in a few years, be produced from a single pair of the most unquestioned purity. Remembering this, we should regard a standard with a sane eye. And for that reason it should be the subject of constant and reasonable discussion, so that its lines may be those of the highest degree of beauty consistent with the fullest development of utility qualities.

A word of caution may be added from Miss Croad herself. In her book on " The Langshan Fowl," page 95, she says :

> " We would earnestly appeal to the members of the Club and Langshan breeders generally, not to groove in too closely for a standard. Not to lose sight of the useful and intrinsic qualities of the breed in order to obtain feather and points of ideal perfection."

If these wise words are allowed to remain the governing principle of the Club and its members, the prosperity and popularity of the breed are assured.

A note may be added as to the best method of breeding either exhibition or utility stock of the highest value. It may be expressed in the following simple rule : Rear as many as possible, but breed from as few as possible. No other method will produce the best and surest results, but by following it closely the breeder will, in the course of a few years, be able to forecast the type of the birds he is breeding as well as to count upon their utility value.

Croad Langshans.

By R. O. Ridley.

Mr. R. O. Ridley.

THEY were first brought to my notice through Mr. Herbert P. Mullens about twenty-five years ago, who had had eggs from the late Miss Croad. On Mr. Mullens having to go abroad on account of health, I purchased his birds, also eggs and birds from the late Miss Croad. Just at that time there was a great paper controversy as to what the original imported Langshan was, and the late Miss Croad fought hard to keep the birds pure, and was much averse to any one using any of her stock for crossing. The late Miss Croad, with the help of Mr. Housman, started a Club called " The Pure Langshan Club " (at the express wish of Miss Croad I consented to be the President), but the breeders of the tall Langshans objected to this, and there were many heated letters in *The Feathered World*. I was always of an opinion there was room for the two types. The name was changed to " The Croad Langshan."

The late Miss Croad came to stay at Heacham, not far from where I live (she was then nearly eighty), and she inspected all my birds and went away quite satisfied that my flock were " pure bred," which I took to mean were like the original birds that she had received from her brother.

The late P. Proud was a great believer in the merits of the Croad Langshans, and he did a lot to bring them before the public by his recommending them in *The Feathered World*. Personally, I have found the breed a very good all-round bird and good layers of good eggs. The colour varies, but I have always set the dark brown eggs, and much prefer them. I find the chickens easy to rear, and my stock has been free from any disease.

There were very few Croad Langshans about at that time, and it was really to back up Miss Croad and Mr. Mullens that I took them. Mr. Housman worked very hard to keep the breed going, and since the death of Miss Croad he has been its great mainstay.

CROAD LANGSHAN PULLET.
Bred by Mr. R. O. RIDLEY.

History of the Langshan Fowl.

By R. Fletcher Housman.

Mr. R. Fletcher Housman.

IN 1872 the late Major Croad brought some very fine specimens of pure Langshans from the far North of China into England and kept them for himself; afterwards he presented them to the late Miss A. C. Croad, of Durrington, Worthing, and they would certainly have become very popular with the public, but Mr. Lewis Wright and Mr. W. B. Tegetmeier proved very bitter opponents of the breed—the former saying it was a Black Cochin, pure and simple. Some years later he had some Langshans in his possession, when he confessed that "the Langshan is a distinct breed and it is an admirable fowl; the skin is not only white, but very thin, and the meat is extremely white and vapid." Before the importation of Langshans there were Black Cochins in England—the colour was of rusty black; they had yellow legs and were dull and clumsy. I used to keep Buff Cochins years ago, and I found them very cowardly or chicken-hearted. I have seen Cochin cocks fighting against small Hamburgh cocks, and the Cochins seemed to have the better of the fight, as they were so large, towering over the small rival, but suddenly they would turn and run away, followed by the other.

I find the pure Langshan is quite distinct from the Cochin, as they are very active and alert, and determined fighters till one is badly beaten by a stronger one. When Miss Croad offered a cup for Langshan classes at Birmingham, the breeders of Black Cochins entered their birds in Langshan classes and carried off the cup and prizes—the all-round judge did not know that the Langshan was a distinct breed, so he went for Cochin type. After the importation of Langshans, the Cochin breeders immediately sought Langshans to cross Black Cochins to improve the colour of the plumage, and the legs of Cochins became black.

The Langshans have been unjustly attacked from 1872 down to 1902. In the latter years Miss Croad was bravely fighting single handed in defence of her Langshans, so I stepped in and fought against her opponents, and Miss Croad retired leaving me to defend the breed. At first Langshans were judged by all-round poultry judges; for some years they gave prizes to birds of large size, condition and short on legs. Members of the Langshan Club were puzzled as to which of the type they should breed their birds up to. I was the first specialist judge at the Dairy Show many years ago, and I kept to one type as far as possible throughout. In 1904 I started the Croad Langshan Club, and we framed the Standard of Excellence; the present members are breeding up to that standard.

There is great culinary merit in Croad Langshans, viz., high average of egg production, and hardihood and handsome carriage. Six years after the importation of Langshans I had a setting of eggs from Miss Croad, twelve as fine eggs as I ever met with. Twelve chicks were hatched from them; they grew up very large birds with jet black scales on shanks and plenty pink between toes; all had pure white toe-nails, and I never saw a black toe-nail among them. I bred them many years. I had eggs from Miss Croad from time to time in order to get fresh blood, and later on she had several importations of Langshans in 1883-4-5. Some of the cockerels had rather too large combs, and some of the pullets' combs fell over the side. I did not like this, but Miss Croad said those pullets whose combs fell over were better layers than those with upright combs, but in my experience I find no difference— small, straight combs are as good layers as those large, shaky combs. There are two reasons why I have added the name " Croad " with Langshan : first, in memory of Miss Croad's gallant defence of the breed ; and second, to be distinct from the Modern Langshan. " Croad Langshan " is of the ancient breed.

Some people complain that Croad Langshans lay some small eggs ; but they should not set small eggs from very young birds, but set fine large eggs, either from early-hatched pullets or hens. Some breeds are good layers of fine eggs, but none of them can approach Croad Langshans in eggs, meat, and hardihood. I had a large young Croad cockerel killed and roasted, with the size and shape on table of a turkey. People who had partaken of it said what delicious and tender turkey, and when we told them it was not turkey but a Croad Langshan fowl they were surprised.

The Chinese in the farthest North of China, in the district of Langshan Hills, considered the Langshan fowl sacred and allied to Turkey, but they meant the flavour is just like the turkey. I remember in Mr. Bush's time, years ago, he won many prizes with Langshans under all-round poultry judges. His birds were, in type, like Dorkings, low on legs, large sweeping tails, with two sickles very long and with very long hangers, beautiful beetle-green sheen throughout ; but I never care for low legs—the legs should be higher to be symmetrical, which is very taking to the eye.

Judging Croad Langshans,

By Edward Cocker.

Mr. Edward Cocker.

THIS is rather a difficult question to describe, as the methods of judges vary considerably in regard to the procedure they adopt to arrive at the merits of the exhibits they are called to adjudicate upon. Some judges only go over these with what one would call "judging by the stick" method. This is done without taking out the exhibits to examine each one in regard to points. It is a rather misleading method, as some exhibitors are a little confused to find their birds placed well up at one show and at another sometimes even cardless. Consequently they jump to the conclusion that the judges have somehow missed the merits of the birds they have exhibited. I have watched various judges at work at shows, and can say, to do thorough justice to the exhibits, one must be prepared to devote a reasonable amount of time to the work. The stick method is alright when one desires to get through a large amount of work, irrespective of the merits of the exhibits, but one cannot say that this is the most satisfactory method to pursue. Take, for instance, a show where one or two judges take all classes, and another show where all, or nearly so, are specialist judges, and from which does one derive the most satisfaction ? Without question at the shows where the birds are judged with the stick and handled individually afterwards. I well remember when I first had the opportunity of judging Croads, I sought advice from one fancier, in particular, who had judged at nearly all the leading shows in the United Kingdom. The advice and lessons I got from him were worth following. He was one that one would describe as a typical artisan fancier, heart and soul in his work, and one most widely sought for his opinion on various breeds. A more conscientious man it would be difficult to find. It did not matter to him who the exhibits belonged to, peer or peasant, all had the same equal chance. A favourite expression of his was : " I judge the birds, not the person who may be the fortunate possessor of the same." He was one most thorough in detail and upright in giving his awards ; a more suitable person to follow one could not desire.

I will try to give a few details of what he taught me, and what various judges in this district endorse, and I may add that we are fortunately placed in this respect as regards poultry judging. His first point was : " Never be in a hurry " ; second, " Do not get flurried, but take matters coolly " ; third, " Concentrate on the work at issue." Then, first take a thorough look at the class from beginning to end, and if one finds a bird or two of outstanding type mark these off for later consideration. One will often find in a large class sometimes up to six or seven extraordinary good ones at the first running through. Then go through again and handle from the beginning and score each bird off for points of excellence. This is not quite as easy to do as it may appear here, and one may have to handle the first winner, and the second and third winners, anywhere up to six or seven times before being able to choose out the right one for the proper place. This will give one a faint idea how many times birds have to be put through what one might call their paces. A point I myself strongly advise is that if one comes across two birds that appear to be of equal merit, and this occurs occasionally, is to leave this class till the birds at issue have settled down somewhat and then place them after a final look. Sometimes one will come across two or three birds equal in points on the scoring sheet ; one may excel in colour, another one in head points, and another one in feet and comb. Then one has to see where the best of these excel in some other deciding feature, viz., condition, eye, or full feathering, and also in being well shown and fully matured or well grown. I have a vivid remembrance of one Dairy Show at which I took on the Croad classes—and I do not remember any other show where I was called so over the coals to explain why I had placed a certain exhibit second instead of first. The birds were equal to look at in points from outside the pens, but when I had taken out both and let the person handle same he upheld my decision. The difference was hardly credible unless one went to the trouble of handling the birds.

Another strong feature I make is this. Although some times I know when I have examined an exhibit that this one will be cardless, I make a note or two where deficient, and then, if the owner should inquire, I am in the position to answer the query involved, and maybe enlighten him for future occasions. All this takes up time, but I have found it time well spent, as a little advice is greatly appreciated by a new beginner. Oftentimes this one has either not the means or the fortunate opportunity of being able to get to the show to see how his, or her, exhibit has fared in this respect, and when one has been elected to take on the task, I think it fair and just that one should be prepared to give the information desired if asked for. Although only a detail or two, it may be a considerable help to the person involved.

And now a word of advice to anyone who may be called upon to take on the classes at any future time. Be punctual and in readiness to begin as soon as the time permits, as, from my experience, if one commences in good time the work can be done in a quarter of the time it would take when people are crowding in to the show. This delay often occurs at the smaller events.

Another point—be consistent in regard to type. Do not choose out various types. Type counts strongly with me, and along with condition and colour of feather will go a long way in carrying off the prizes. It takes a lot of the other points to recover the ground lost in this respect. The final word I would give is : Try to get to some show or other where the Croad classes are being handled by a Croad judge ; one can learn in half an hour what may take several seasons to grasp by reading and other methods. I have found from experience that usually the judge appointed will gladly give the information asked for, and fanciers, as a whole, are nearly always open to help others along the road to success.

In conclusion, one will be inclined to say—the foregoing is all right, but how can one arrive at any decision if one should be called upon to adjudicate at some future time ? Well, I will try to give the method and procedure I myself adopt to get at the final result. We will take a class as an example—to take just one individual bird would hardly be a fair comparison.

First, one examines the class most particularly for type and size—especially type—and any other outstanding point of merit, and marks these off accordingly with a cross or other distinctive mark.

Then comes the real test, and one takes out each exhibit and scores these off for points of other merits not seen without having been taken out and examined—such as good colour. Mark off colour in particular; this may be later on the deciding factor in regard to the winners as to where they are eventually placed; condition also mark. This applies to exhibits as to condition shown, and well grown and finished off. At the same time examine for fineness of comb and shape of same; also for colour of eyes; also feet and nails. If any come up to expectations, mark them off with an extra cross, or two crosses, according to their merits. Now one has got a cross each for type, for colour, for condition, and for comb and feet features—say four crosses in all. On arriving at this stage one will find very few have emerged successfully— may be about four or five. Go through these again to choose out the best; and to make sure score off all at the time of handling to save time later on. On going through the last lot as to which may be the best, examine for defects, such as faulty comb, slight blemishes on soles of feet, or purple tinge in feathers—a minor defect one would say. Just so, yet enough to allow the others to score accordingly. This may now only leave two, or at the most three, for the first place, and matters should be more simplified, so that we can decide which has to have the premier honour. If two appear to be exactly equal, look to see which of them is most fully developed in regard to feather and evenness of width throughout when looking at the bird from the front of pen. This sometimes will be the deciding factor as to which approaches the ideal.

I have not said anything about defects, as one marks them down as one judges the birds, and if an exhibit should fail in a minor defect it takes a lot of excellence to compensate for what it may have lost. Five or ten points may not appear to be a great lot, but it proves so when others may excel in other respects.

The Croad Langshan Mainly from the Utility Side.

PART I.—HISTORY AND DEVELOPMENT.

By ALEXANDER SMITH.

MR. ALEXANDER SMITH.

"THE heathen Langshan is peculiar." The words are those of its bitter opponent, Mr. Lewis Wright, and many a truth has been uttered in jest. The Langshan has the size and quality of flesh of the "table" fowl, and yet has a fineness of bone which arrests the attention of the egg-production expert. We had thought this subject would have fallen to the pen of a practical poultry farmer, and would willingly show our appreciation of the compliment implied in the Editor's "Will you say something on 'Utility'?" "Utility" is, fortunately, a word of wide connotation, and many of its aspects are treated so fully in the poultry papers by persons with wide experience that they need hardly be mentioned in such a work as this. We are all concerned with utility: the man who cares only for exhibition points does not interest himself in the Croad Langshan, for, verily, it is peculiar.

From a utility point of view, we start with several advantages. Croad Langshans are of one colour, have one type of comb, and require no system of double mating. They can therefore be bred for utility without the introduction of complications which would prevent the exhibition breeder from availing himself of the resulting improvement.

While the Croad has never been held in high regard as a show bird, and so has escaped the fate of some other breeds, neither has any attempt been made, on the other hand, to convert it into an " egg machine " at the expense of its other utility qualities. It has assuredly been greatly to the advantage of the breed as a whole that it has not been split up, as some others have been, into separate and distinct " exhibition " and " utility " strains.

The writer took up Croad Langshans in 1910 with the intention of breeding them by trap-nest for utility qualities—chiefly laying. In order that he might have as little as possible to " breed out," he went to what he imagined to be an " exhibition " source. He knew nothing about fowls, but as fowls were to be kept he procured one or two poultry books and a sheet of squared paper. Down the left-hand side went the names of all the breeds mentioned. Across the top of the columns went the headings : " Sitters, Size, Table qualities, Layers, Winter layers, Egg size, Egg colour, Hardy, Suit clay," etc. When the books were read, and the notes examined, the Langshan came out first, with Sussex as reserve. It must be said here, however, that considerations of latitude, altitude, and exposure were important. That he had never seen a fowl of either of these breeds did not matter. Evidently there were two kinds of Langshans. He wrote the Presidents of the Langshan Society and the Croad Langshan Club, asking the price of their best eggs, and what was the difference between the " Modern " and " Croad " ! The late Mr. Harry Wallis, who, by the way, kept both types, and won at the Dairy Show the next year with a Croad, ignored the second question. Mr. R. O. Ridley strongly disadvised the keeping of Croads if the object in view was exhibiting, saying they were very troublesome to get to come to colour and type, but if the object were utility the Croad would take a lot of beating. The " Modern " Langshan was of similar character, but had been more carefully bred for exhibition points. He could not speak of them from experience, for he had never kept them, " not liking the look of them."

The four pullets hatched from the first sitting were trap-nested, and put up an average of 176·7 eggs in their first laying year. This altered the question. It was beyond what might reasonably be expected of so large a fowl, as will be better understood from a later part of this chapter. It was not necessary to concentrate on laying, and more attention could be given to other qualities.

In the course of ten years' careful breeding with trap-nests, toe-punching all chickens so that their pedigree could be traced, there has been opportunity for acquiring a sort of general knowledge of the main principles which enable one to breed for whatever purpose one has in view with a minimum of culling. A breeder's object is to secure uniformity rather than the production of exceptional individuals. The ten years' breeding has not produced any layer excelling the best of the original birds, nor has their average been exceeded, but whereas in the first few years there was extreme variation in laying, in height on leg, in amount of leg feather, in shape, size and colour of egg, and so forth, leading to much culling, it is now unnecessary to breed more than a very few to replenish the stock, since little culling is required, uniformity of laying, etc., being marked.

The following are some particulars we noted regarding birds from a hatch of eggs from Mr. R. O. Ridley in 1910.

From 12 eggs 11 chickens were hatched on April 1st. Six cockerels and four pullets were reared: the other was found dead, a weasel being seen. *They were reared on clean ground.*

		Weights.	
COCKERELS:	At 16 Weeks.	At 19½ Weeks.	At 7½ Months.
No. 1 ..	4⅞ lbs.	6⅛ lbs.	8 lbs.
,, 2 ..	5 5/16 ,,	7 ,,	9¼ ,,
,, 3 ..	5⅝ ,,	7⅛ ,,	9 ,,
,, 4 ..	6 ,,	8 ,,	–
,, 5 ..	5¼ ,,	6⅝ ,,	8 ,,
,, 6 ..	5¼ ,,	6⅜ ,,	–
PULLETS:			
No. 1 ..	4⅜ lbs.	5¼ lbs.	–
,, 2 ..	4⅜ ,,	5¼ ,,	–
,, 3 ..	3⅜ ,,	4 ,,	–
,, 4 ..	4⅝ ,,	5⅜ ,,	–
Average, Cockerels	5⅜ lbs.	6⅞ lbs.	8 9/16 lbs.
Pullets	4 3/16 ,,	5 ,,	–

Particulars of laying of above pullets up to November 30th, 1910:

Pullet.	Commenced.	Age.	Eggs.	Days under observation.	Most eggs laid in succession.	Longest missed laying.
No. 1	August 31st ..	5 mo.	75	92	22	3 days
,, 2	September 21st	5⅔ ,,	49*	63*	6	2 days
,, 3	September 22nd	5⅔ ,,	60	70	20	1 day
,, 4	September 22nd	5⅔ ,,	56	70	30	3 days

* Time broken by exhibiting.

The eggs averaged 10 to the lb. at first, and by the end of November, 8 to the lb.

No. 1 laid 219 eggs in the year, and the average for the four was 176·7.

There are three types of fowl—the " Laying " type, the " General Purposes " type, and the " Table " or " Meat " type. What are the limits of expectation of eggs from the various types may be shown by figures from a table of Mr. Walter Hogan's, dealing with the Hogan system of selection. The type depends on the thickness of the pelvic bones, the capacity on the depth from the pelvic bones to the point of the keel. Taking a five-finger (3¾") abdominal capacity—i.e., an extremely large capacity—Mr. Hogan considers that birds of the three types showing this capacity may be expected to lay as under :

The *Typical Egg Fowl*,   from 175 (⅜" pelvic bone) to 250 (₁⅛").

The *Dual Purpose Fowl*, from 85 (¾"   „   „) to 160 (₁₇₆").

The " *Meat* " *Fowl*,   from   0 (1⅛"   „   „) to 70 (1³").

From this it will be seen that the four pullets already mentioned averaged more than the expectation for birds of the best " capacity " of the Dual Purpose type, to which the Langshan belongs.   The fact is that in the Langshan we have a bird differing somewhat in skeletal structure from other fowls of its type. Against it in measuring for abdominal capacity we have that the keel is rather long and begins far back, hence the point of it comes rather far to the rear, but on the other hand the bones of the framework are remarkably thin for the size of the bird, and their thinness places it higher in the laying scale than might be suspected from the capacity measurement alone.

This leads naturally to the subject of Laying Contests. A veil of silence has shrouded the performance of three pens of Croads which took part in one of the earliest laying competitions at the Harper-Adams College. Yes, one of them was the writer's. Now that Miss A. W. Simmons' pen of Croads at Bentley in the N.U.P.S. National Egg-laying Test is in a fair way to recovering lost laurels, the veil may be lifted for the purpose of discovering the lessons to be learned. Of the six pullets sent, four were daughters of a hen which laid 219 eggs in her first laying year, and 238 in her best laying year. The change of climate, feeding, etc., evidently upset them and caused a partial moult, for it was more than three months before the pen made a regular start. The four sisters showed great diversity, one of them laying a number of eggs barely exceeding half of the average made by the others. This extreme variation suggested that the mating of their dam had not been suitable, or that she herself was the result of an out-cross of strain, and that her high fecundity owed something to exceptional stamina. This seemed probable, for extreme variation was found in the " build " of her offspring after three different male birds. She herself was practically devoid of leg and foot feather, but her offspring varied greatly in this respect, and in height on leg. Mr. Ridley astonished us by pointing out all her chickens without a mistake in a considerable flock before he had seen the hen herself. This hen was very useful, as she bred show birds. Our recollection is that two cockerels which stood third and V.H.C. at the Crystal Palace were the full brothers of these four pullets. One of the pullets sent to the competition had a " Best pullet in Show " to her credit before she went up, at a local show.

This part of our records, both of laying and showing, passed out of our hands for scientific purposes some years ago, but there is an illustration here of two things ; type and high laying powers were found together, which speaks well for the exhibition standard, and exhibition birds of both sexes were bred, not only from one pen, but from the same dam. As the local show is no criterion, it may be well to add that the pullet in question, after her return from the competition, was sent to Mr. Housman for breeding purposes, and in his hands won at least two third prizes under club judges. Her sons found their way to other breeders, so, although we later found we had made a mistake in breeding from her female progeny instead of the males, as will appear later, the advantage was not lost to the breed. The need for uniformity had been learned ; another conclusion reached was that if birds were to be sent to a laying competition again they would have to be fed rather more poorly than we feed in this cold climate, so that they would find themselves in better and more encouraging conditions on transfer ; and a third lesson, which the promoters of competitions were quick to learn, was that it is unreasonable to put all types of birds in the same class.

In the National Utility Poultry Society's Egg-Laying Test presently in progress at Bentley, there are six sections. The first three are for the three varieties which have been most carefully bred for egg-production, viz., White Leghorns, White Wyandottes, and Rhode Island Reds. Section IV. is for Sitting Breeds (other than the above) ; Section V., Non-sitting Breeds (other than White Leghorns) ; Section VI., is the " Championship " Section. It is in Section IV. that we are chiefly interested. There are twenty-four pens competing—nine of Buff Orpingtons, seven of Light Sussex, three of Speckled Sussex, three of Buff Rocks, one of White Orpingtons, and one of Croad Langshans.

The Croad Langshans early took the leading position, and the report of the tenth month shows them still in the lead, with a test score value of 729, the next in the section being 685. Miss A. W. Simmons, to whom the pen belongs, has been good enough to accede to a request for particulars of how these birds were bred and managed. She writes :

" The original pen of six one-year-old birds came from Mr. F. Joergens, of Reading, in 1916. They were mated to a cockerel from a sitting of eggs from the same breeder, though as to their relationship I cannot say. They laid very well, but my sister, who had them before I took over the poultry, did not keep a record. I took it over in October, 1916 ; no fresh strain was introduced until November, 1918, when I had a cockerel from Messrs. Whitaker and Tootill. I ran him with nine hens that had done well in their first season, and the birds at Bentley are from that mating.

" The birds in the competition were hatched between January 22nd and April 28th, 1919. Two of the early-hatched ones laid the fourth week in June, and three others joined in within a month. They laid steadily for a few weeks, then brooded and moulted till October 12th, when two laid again. The totals for those weeks from June 28th to August 23rd were as follows : 12, 18, 28, 29, 22, 19, 11 (equals 139) from the five birds. . . . I finally sent one of those five and four sisters, hatched probably March and April.

" . . . The birds from which Pen 213 was bred had the run óf the meadow, but hardly any corn at all ; I could not get it, and used vegetables (carrot, parsnip, turnip, beetroot and potatoes) pulped, sometimes cooked, mixed with bran, meat meal, maize meal and toppings. I used raw vegetables for mornings to give the digestion something to do, and cooked at night as soft mash, with clover hay.

" Last summer I could get scarcely any bran or toppings from June to October, and had to use maize meal and grocers' waste (usually macaroni, rolled oats and maize), consequently the birds were all too fat, and I think that was partly the cause of the pullets maturing early. I use butter-milk and skim for young birds and chicks.

" I like a few pullets to mature early ; they give some idea as to the laying powers of the season's hatching, and I have not found them any the worse for it.

" . . . *Re* the size of the egg. I weigh only those that look less than 2 ozs., but I think I should be within the mark to put their average at $2\frac{1}{4}$ ozs. The colours range from a rich brown to a shade that is almost pink. I find turnips improve the size of the eggs, also mangels. . . .

" I have not exhibited at all, but I think my birds are all a nice type and a lovely green. . . .

" I think the birds went to Bentley the last week in October ; the test started November 7th."

Miss Simmons has twenty-one other pullets bred from the same pen as Pen 213, and greatly enhances the value of the test score by being able to give for comparison the weekly number of eggs laid by these. The figures are for the twenty-seven weeks beginning with the week ending November 8th, 1919, and finishing with the week ending May 8th, 1920. They are : 7, 13, 31, 45, 55, 56, 45, 29, 27, 27, 35, 41, 40, 44, 62, 56, 79, 82, 103, 101, 103, 80, 82, 73, 63, 64, 41—total 1,484, for twenty-one birds. In this connection it should be mentioned that the test score of Pen 213 already given happens to be identical with the number of eggs laid.

Laying competitions have served a useful purpose in drawing the attention of the public to the fact that by selective breeding birds of high and reliable fecundity can be evolved. Too often, however, the mistake is made of giving the credit to the *breed* and not to the *strain*. For success, much undoubtedly depends on the ability of the owner to select the most likely layers among his pullets, and on the feeding and management of the birds before they go up. The less change of conditions they experience on removal to the venue of the contest the better : change of altitude seems to affect them more markedly than other changes. The first two-year competition, held in Australia, was won by a pen of Australian Black Langshans, which are of the Croad type. Their

"ROMULUS."
Bred by MR. R. F. HOUSMAN.

average of 414 eggs per bird established a world's record. **Mr.**
Bryant imported birds bred from this pen, and considerable use
was made of them.

Mere egg-laying competitions, however, fail to give information on points that are of vital importance from a commercial
point of view. Too often the high fecundity has been achieved
at the expense of other qualities of equal importance. The birds
winning these competitions were often very small, but this must
not be taken as implying that under-sized birds are better layers
than those of medium size for their breed : such is not the case.
Their eggs too frequently gave poor hatching results, and too often
chicks bred from them were difficult to rear. Too many of their
eggs were small. These points are now studied, and safeguards
are introduced in the rules governing the scoring. It is desirable
here to caution the reader against a too hasty assumption that high
fecundity is necessarily accompanied by difficulties of hatching
and lowered vitality of the chickens. When one breeds for the
abnormal, whether in respect of number of eggs, size of egg, size of
bird, or indeed anything abnormal, a lowering of the reproductive
powers and vitality may be expected, and intelligent precautions
must be taken to guard against it. In the ordinary course of
nature the birds whose offspring have most chance of surviving
are those with most stamina. The breeder is, for a special purpose,
endeavouring to secure the survival of birds with certain special
characteristics. His success in this, if he fails to realise that the
continuance of the race depends primarily on stamina, may be
rapid, like the growth of the gourd, but will certainly be as transient.
Trap-nest records are extremely useful, nay, essential, but they
are traps for the unwary. Let every user of trap-nests, and the
larger number who pin their faith to records, ponder well, and try
to grasp the full significance of the statement of a well-known
breeder that the egg-record indicates, not the breeding value of the
bird, but the breeding value of her parents, or rather of her
ancestors.

The use of trap-nests, with the recording of the individual
scores, entails much extra labour. However little knowledge of
breeding the user possesses he cannot fail to reap advantage from
the elimination of bad layers, but the full value cannot be got out
of their use unless some study of the principles of breeding accom-
panies it, and one knows how to mate the birds to best advantage.

Some of the points to be considered in the selection of a good pattern of trap-nest may be mentioned.

(i.) The action must be reliable : it must be impossible for a hen to enter the nest, lay in it, and leave it without being " trapped."

(ii.) There should be as little mechanism as possible.

(iii.) It is a great advantage if there is a perfectly plain floor to the nest, with no obstruction to interfere with cleaning out.

(iv.) Some drop-shutter nests require head-room : they cannot be placed under dropping-boards, and yet have to be placed so that fowls cannot perch on the top of them.

(v.) A falling door on hinges is better to be made so that it can be opened outwards as well as inwards, to avoid forcing an excited bird back upon her egg.

(vi.) The action must be capable of being " set " by one hand, and if it is self-setting, so much the better.

A trap-nest intended for Langshans must be of more than usual size. The Langshan likes to have room for her tail when she stands up to lay, and this is the reason why some people find this breed will not lay in their trap-nests. If they are given a nest sufficiently high they prefer a trap-nest to any open nest. The amount of noise made by the shutter is a consideration with nervous breeds, but not in the case of Langshans.

The nests we have in use are of our own making, and are of two patterns. One is on the drop-shutter principle, adapted from the design of a friend. The shutter, of three-ply wood, is held at any height by thrust and friction, and an ounce or less of pressure on the lever is enough to release it, while we have seen a heavy cock perching on the top of the shutter, his weight only serving to hold it more tightly. Had a hen gone in, —— (?). This nest works well, but fails in respect of (iii.) and (iv.). The later pattern is one of which a sketch was published in *The Feathered World* some years ago. No explanation was given, except that it was a pattern in use at an American laying competition. It is of the hinged falling-door type, the catch is simply a shaped block of wood turning easily on a nail at the side of the nest, and there are no springs, wires, or anything else to get out of order. It is difficult to imagine anything simpler. The door easily opens outwards, the lower board in front unhooks, and the floor of the nest, and five inches above it, are clear of any obstruction. It is not self-setting, but can be set with an easy motion of one hand. We have not seen anything on the market which seemed to possess all the advantages of this nest. The shutter, whatever its pattern, may, with advantage, but only if the nest be large, admit light at or near the top. This encourages the hen to rise and demand to be released, it allows the attendant to look in, and if the hen is restive she is misled into trying to get out at the top, where her ingenuity is wasted. One other point should receive attention if the nest is home-made. Ask yourself whether, in the event of the hen's succeeding in getting her head under the shutter or door and finding further progress impossible, she may not hang herself on trying to withdraw her head.

Personally, we have come to pay less and less heed to the actual score recorded, and regard the discovery of exceptional layers as a minor function of the trap-nest. Its main uses are to weed out the bad, and, in conjunction with the toe-punch, to discover the best breeders among the better layers.

The toe-punch is of use in other ways. It enables one to trace to its source any sign of alien blood which may show in chickens. Should a chicken appear with, say, black feet, one can at once locate the sire and dam. An examination of the other chickens of this hen, and of those of the sire by other hens, will soon disclose to the suspicious eye which parent has the taint. The user of the toe-punch can then proceed with certainty to eliminate all the progeny of the offender, while the breeder who does not use it is likely to get deeper into trouble through using some of them which do not happen to give outward indication of cross blood. It also leads to the discovery of any case of what is called a " fortunate nick," greatly to the breeder's advantage subsequently.

The toe-punch is an instrument which cuts a tiny hole in the web between the toes. The first position is between the outer toes of the right foot : in other words the four webs are reckoned in order from your left when the bird is facing you. One, two, three, or four marks may be used, or no mark at all. It may be as well to detail the different possible markings, which are : 1 ; 2 ; 3 ; 4 ; 1, 2 ; 1, 3 ; 1, 4 ; 2, 3 ; 2, 4 ; 3, 4 ; 1, 2, 3 ; 1, 2, 4 ; 1, 3, 4 ; 2, 3, 4 ; 1, 2, 3, 4 ; no mark—sixteen in all. It is best to avoid having any chick with no mark, as the web may get torn, or the tiny disc punched out may not be detached, with the result that the hole may close, in which case the bird may be confused with those that have no mark. The two feet of the bird should be compared if there is any reason to think that a punch-mark has disappeared. The hole grows larger as the bird grows, and it should be made well back in the web. The marks of the toe-punch are expressly exempted from disqualifying the bird from exhibition. The pain inflicted on the newly-hatched chick is momentary, and if any lady reader feels compunction let her consider that the mark indicating good pedigree will tend towards ensuring the chick a longer and happier life. In the case of Langshans, marks in the second and third positions are to be preferred for most frequent use, since the foot-feather often conceals the first and fourth, making it necessary to handle the bird to identify.

The birds that breed the best layers are those that lay well in winter : this point is of the highest importance. It is not the year's record, but the record for the winter months which shows the probable breeder of good layers. Now the vital point is this, that in spring, just when eggs are being used for hatching, these are the very birds which are not laying freely. As one writer phrased it recently : " Every bankrupt ' duffer ' of a hen " is laying at that time, and unless care is exercised the result will obviously be the replenishing of the stock mainly from the worst layers !

Where trap-nests cannot be used the best breeders may be detected, though with less certainty, in this way. Rings of three different colours being provided, all birds found on the nest " on business " in November are rung with one colour ; the colour is changed in December, and again in January—a second and third ring being given to the same bird if still laying.

Birds wearing all three of these colours should be selected to be bred from. If more are needed, those with the two colours for December and January are to be preferred, for it is better to use a bird that has made a late start than one that has stopped for a whole month after she has once started.

Housing needs little comment. A good house is an advantage, but the Langshan is one of the hardiest of fowls, and if the house has plenty of fresh air admitted will do well.

Feeding is a matter of great importance, but this subject will be found exhaustively treated in all poultry books and magazines.

Perhaps it may sound incredible, but it is nevertheless true, that the writer does not know either the constituents or the albuminoid ratio of the prepared dry mashes on which his birds are fed, nor does he even care to inquire. This, be it said, does not arise from ignorance of such matters, or failure to realise their supreme importance, but rather from the fact that, having at one time made a study of advanced organic chemistry, among other things, he is convinced that the subject of feeding requires much more study and specialised knowledge than he can profitably bring to bear on it. He is content to let others experiment on *their* birds, and to save trouble by adopting what they have proved to give good results for several years. Methods of feeding suitable to the colder climate of the north of Scotland would produce other effects in England, so the subject may be dismissed with a few remarks on the wet mash and dry mash systems.

Wet mash is probably best for giving size, and for growing chickens quickly ; it is more palatable, and they eat more. House scraps can be used up, and meals purchased in small quantities, so as to be fresher. There is greater variety, which tends to promote growth, and where labour has not to be paid for it is probably the more economical system.

The dry mash system of feeding saves a great deal of labour, a consideration of the first importance on a commercial egg farm, as distinct from a poultry farm. A well-balanced ration can be supplied with less trouble, and there is less risk of over-feeding. Fowls prefer it freshly mixed, or well aerated. If you do not mix it yourself it is advisable to have at least two different mashes, to provide a change of diet. A good pattern of hopper is essential. Magazine hoppers, even of the best pattern, are somewhat wasteful, and bad patterns are vexatiously so. The box hopper needs more frequent filling, but as it does not choke like the magazine type, it does not need the same amount of attention, or rather does not need to be attended to so often ; it is cheap and satisfactory, being less wasteful. A really good magazine hopper has, of course, certain advantages. An excellent pattern of box hopper which is very easily made is described and illustrated in the Annual Register for 1918 of the Scientific Poultry-Breeders' Association. Our birds have been changed on to dry mash recently, and by the expedient of leaving the hoppers open and reducing the wet mash gradually the change was effected without any check to the laying

The chickens, however, are given a wet mash last thing at night—usually just the dry mash scalded. If the hens are to be given scraps, or a warm wet mash in winter, it should not be given them till far on in the day, if dry mash is to be their staple food, otherwise they will neglect their dry mash in constant expectation of something more tasty.

Breeding stock give better results when a larger proportion of hard feed is used.

Langshan chickens are very easy to rear. They need no special treatment, and indeed no special remark except that they should not be kept in coop or other shelter of a close nature on account of bad weather, if it is at all possible to let them out. They will stand very severe weather, but they will quickly die off if shut in. Any common-sense feeding, with suitable grit and chopped green food right from the shell, should answer. Boiled rice fed occasionally helps to prevent diarrhœa, but it is not advisable to give much rice to chicks if size is desired. A little Parrish's Food in their drinking-water will do them no harm.

Where hatching is begun very early, for exhibition purposes or to get early cockerels for breeding, it may be found necessary to have chicks indoors for a time if there is a storm of long duration. Should this be the case they need come to no harm if attention is paid to two things. The first is to see that they have green food, the second is to let them have access to damp (not wet) earth under their feet from the twelfth to the twentieth day after they were hatched, i.e., during their third week. If they are with a hen she will take them on to the damp earth and sit down on them. They should not be on the damp earth at night. A large amount of bran in their food is good for chickens.

Some people do not care for Langshans because of their broodiness. Let us admit, right away, that some Langshans could do with a good deal less of this, though perhaps we would not go quite so far as to agree with the gentleman who expressed to Mr. Ridley his conviction that some of them would hatch door-knobs ! Our Langshans had to yield the palm in this respect to a farm broody we had sitting on early eggs a year or two ago. She was relieved of her charge and returned to the farm about the twentieth day, as she took the chickens out of the shell as soon as they made a noise. Later she passed into the possession of a game-keeper when the pheasant-hatching season came on, and disclosed her identity by repeating the performance. The keeper, however, simply transferred her eggs to another hen, and gave her a fresh lot. When last we heard of her she was sitting out the third successive lot of pheasants' eggs. It is generally well into the spring before the Langshan goes broody. Though our birds were in lay all winter we had nearly always to look elsewhere for broodies for our early hatching. Our egg records also show that if a hen or pullet is "broken off" at once she lays again about ten days later. The second time she goes broody she takes a fortnight, and the third time in a season she needs three weeks to resume. By that time it is getting on for the moulting season, and it is not a bad plan to let her sit till the moult is begun, so as to get her moulted early.

But there is another side to the question of broodiness. Many, especially those with small stocks, rely on the natural method of incubation, and nothing beats a Langshan for this purpose—she is reliable, quiet, easily handled, and most careful of her eggs and chicks. At first we feared that so heavy a bird would be clumsy, and that the foot-feather would entangle the chicks ; it was quite a mistake, and for a valuable sitting of eggs give us a Croad—even a pullet—in preference to any other. An American who came to this country and set up an intensive egg farm used a broody breed —Buff Rocks, if our recollection serves. This fact was commented upon by a visitor with some astonishment, but the reply of the owner was that he preferred a bird that " *tells* you when she is not laying," for you can then take steps to get her into lay again, whereas the non-sitting breeds, in their idle time, " go about looking as if they were laying all the eggs on the place."

Broodiness is hereditary, and it lies with the breeder to keep it in moderation by refraining from using the sons of very broody hens : like most other characteristics, broodiness is transmitted chiefly through the sons. After all, our records do not show that the non-broody ones—few in number—or those seldom broody, excelled the others in egg-production. The best layer we ever had was five times broody in her first laying year. A very broody bird generally lays very freely between her spells of broodiness.

A bird of this habit, however, does not breed so good layers as one that lays the same number of eggs but lays them fewer at a time and with shorter intervals between.

An amusing illustration of maternal instinct was given us not long ago. A third year breeding hen had had a wound, and, being broody, was accommodated in an open trap-nest on the floor level. Some corn was put in the nest for her, and from time to time she was heard trying to attract some well-grown chickens which were using the house. In this she had no success, but later in the evening, when we looked in, her behaviour and the sounds she made were like those of a hen with chickens. It occurred to us that she might have left the nest, and another hen taken her place, so we proceeded to raise her breast in order to see her ring. This she strongly resented, but when she was forced up she was found to be keeping in custody a half-grown sparrow, which blinked at the light, and promptly flew away.

A pullet ought to be well matured before she commences to lay. Seven months is held to be the proper age for this breed, the Langshan being a large bird and requiring rather longer time for development in consequence. We have found the average to be just six months ; at first there was variation of from $4\frac{1}{2}$ to 8 months. If pullets can be kept back till seven months or longer it has a good effect on the ultimate size of the egg. Of the pullets of one hatch, reared together, the first to commence laying is nearly always the best layer. It is a sign that you are getting uniformity in breeding when all the pullets of one hatch come on to lay within a few days of each other ; when such is the case you may take it that there will be very little difference in their year's records. Against the slow development of the Langshan one must put this other fact, that she is good for more years as a layer. Indeed, some very old hens continue to lay well, but that is not what is wanted from a utility point of view. The most profitable bird is the one which lays plenty of eggs in the first two years : these are usually the average-sized ones which come on to lay in normal time. Still, it is worth consideration that a bird that takes longer to develop, like the Langshan, retains its useful qualities for a greater length of time. Two farmers in this county were conversing about a servant who had, some time before, been in the service of one of them. The other had engaged him, and asked his friend's opinion of him. The man's former employer gave him a good character, adding that the only fault he had to find with him was that it was difficult to get him up in the morning. The other's rejoinder was : " The thing that interests me is not so much what time he gets up as what he is worth when he *is* up." The Langshan will stand this farmer's test.

Let us now turn to the egg. As a breed the Langshan ought to lay a large, dark brown egg. Different individual hens differ widely in colour of egg : not only so, but while certain hens lay eggs very uniform in colour, others show extraordinary diversity. Once we had four or five successive eggs from one hen as different in colour as we could have found among the rest in all our experience —white, brick-red (without either polish or bloom), plum-colour with a whitish bloom, and pinky-brown, which is the desired colour. It was frequently found that birds of very good exhibition type laid light-coloured eggs, which led to some speculation as to whether the dark egg was not a later development. Some associated the dark egg with hard metallic green sheen, very dark eye, dark beak, and black in toe-nails. A gentleman who kept Langshans when resident in China informed us that the eggs of the Chinese birds were decidedly large and dark, and our own experience in breeding for dark eggs has been that we have had fewer than before showing any black in nails, some with almost white beaks, and some with light eyes, while we have lost somewhat in sheen. This does not show that dark colour of egg goes with light beak, white nails, etc., for we have been selecting for these things as well, but it goes to prove that it is independent of the dark beak, etc. There were several other ideas about the dark egg which seem to have as little foundation. It was thought that they were, on an average, smaller than others. There is probably a slight optical illusion here : a rifleman will tell you that a dark object looks smaller than a light-coloured one of the same size. Selecting brown eggs for a show we can generally get them to average 2·4 ounces, or a little more ; the heaviest dozen we remember having picked for colour averaged 2·52 ounces. Another idea was that they more frequently failed to hatch. This seemed not unlikely, since the selection was for the abnormal, but experience does not bear it out.

The average egg of the adult Langshan, as far as we have seen, is between $2\frac{1}{8}$ and $2\frac{1}{4}$ ounces. It is certainly possible to breed a strain laying eggs approaching $2\frac{1}{2}$ ounces, with a goodly number exceeding that weight, but except for exhibition purposes it is not worth while, since the disabilities attaching to breeding for the abnormal make it difficult to maintain. An egg weighing $2\frac{1}{4}$ ounces is quite large enough. For exhibition purposes the egg should be large ($2\frac{1}{4}$ to $2\frac{1}{2}$ ounces), smooth, evenly coloured all over, free from spots, and as round as possible, so that one can hardly tell which

CHAMPION PULLET, 1911.
Property of Mr. E. COCKER.

is the " big end." Those who think of exhibiting eggs would do well to grasp that the chief considerations are freshness and uniformity : an egg a little darker or larger than the rest spoils the appearance of the lot. Double-yolked eggs are not laid by the different hens indiscriminately, but are due to a comparatively small number of hens which seem to make a speciality of this. Popular belief that the hen which lays a double-yolked egg is about to stop laying, or will not lay next day, is mistaken. Such eggs hardly ever occur when the bird is stopping, and though a day is generally missed it is usually the day before.

There are two kinds of brown eggs. Some have a whitish shell with a brown surface-colour ; this can be rubbed off with a wet cloth, if a little time be spent on it. These eggs often have a very dark appearance. The other kind have a brown shell, often with a light bloom on the surface which makes them look quite light in colour. On being wetted these are seen to be very dark, but the light bloom does not rub off, and on drying they resume their original appearance. The latter are the ones to use in breeding for egg-colour, in preference to the former.

The practice of " resting " eggs which have come a distance is one for which there is no supporting evidence. At best the good done is problematical, while the effect of adding twenty-four hours to the staleness of all the eggs is certain.

Beginners frequently err in setting too small a number of eggs for their requirements. Disappointment might be avoided if they knew that experienced poultry farmers set five eggs for every pullet they want to rear to the laying stage ; if they have one laying pullet for every four eggs set they consider themselves lucky.

Fertility is a thing which can to a greater extent than might be imagined be controlled by breeding and it can be influenced by feeding. The condition of the male bird is of importance. He should be given a feed by himself, apart from his hens, at least once a day, and *all he can eat*. Keep the tit-bits for him, and do not allow him to give them to the hens. The best results in hatching are got when the cock is in good condition and the hens in rather poor condition. If strong winds prevail fertility tends to be low, as the cock is apt to stand about in shelter and pay little attention to his hens. The number of hens to a cock depends on his individual amativeness, his age, and whether the birds are on free range or not. A shy cockerel which never pays attention to his hens when anyone is about is generally found to give very high fertility. Seven hens to a cockerel or young cock is usually plenty, if they are kept semi-extensively. If a fresh male bird is put in the pen, we have known eggs laid two days later to have been fertilised by the new male : by five days most of them will be after him, but twelve days must be allowed to elapse before you can be certain that all the fertile eggs are after the new male.

The hatching egg should not be deficient in phosphorus. The theory, due, we think, to Bunsen, is that during incubation the phosphorus in the yolk becomes converted into phosphoric acid, which attacks the shell, forming phosphate of lime, serving the double purpose of weakening the shell and hardening the bones of the chick. Dead-in-shell chicks have rather soft bones.

All eggs being incubated should be tested. The removal of the colder infertile eggs gives the others a far better chance, and the eggs removed can be fed to chickens. If there is a bad egg or two in the nest at hatching-time the effect of the bad air is harmful. With dark brown eggs it is best to delay testing till the ninth or tenth day; testing can then be carried out with speed and certainty. A point worth remembering is that it is from the fifth to the seventh day that the embryo is most easily killed, and for that reason alone it is at that stage well to avoid the disturbing effect of unnecessary cooling, and of the strong light used for testing. We do our testing by means of an acetylene cycle lamp and a piece of cardboard or wood with a hole the shape of, and slightly smaller than, the egg.

What about the sex of the chickens ? We have no experience of modern instruments, and have found nothing to confirm old wives' theories. Research shows no evidence in favour of the idea we once accepted that early hatches give mainly cockerels, late ones mainly pullets. It may be so, but if it is, it is not due to a larger proportion of early eggs containing cockerels, but might be due to the cockerels being stronger and so more " hatchable " early. Some hens are cockerel-breeders, and others pullet-breeders, in a literal sense, although most hens breed about equal numbers of either sex. This is hereditary, the tendency to breed a majority of one sex being transmissible from dam to son, and probably in a less degree to daughter. Here again the breeder can exercise some control, but the breeder is having rather many things to control ! Still, he will consider this point among others, if he discovers a hen with this tendency. The theory is that only one of the two ovaries is active, and that the two sexes are produced from the two lobes of the active ovary : that frequently one lobe is more active than the other, and hence a majority of one sex results. But where does the male come in ? According to this theory he has no influence on the sex of his immediate progeny, but he can transmit to them the tendency to breed one sex if he has inherited such from his sire and dam, particularly the latter. Any reader who is particularly interested in these theories may be referred to the work of the late Mr. Oscar Smart, formerly the official adviser of the Scientific Poultry-Breeders' Association. We had some correspondence with him, being somewhat sceptical, but it turned out that our very full records, when we examined them in the light of these theories, verified them in directions we had never suspected. This interested him : he made an examination of our records of Croad Langshans, and expressed a high opinion of their powers and possibilities as layers, but his failing health prevented him from following the matter up as he meant to do. An interesting experiment is being carried out at present to see how far it is possible to determine, early in the period of incubation, the sex of the embryo from the amount of carbon dioxide given off. The proportion seems to be more than double in the case of cockerels, but the method is not capable at present of being applied on a commercial scale.

Some other points regarding hatching eggs may be of interest. Eggs stored upright in a rack keep all right without being turned. We frequently set a whole sitting from one bird, taking up to three weeks to collect. Eggs more than a week old are usually useless for artificial incubation, but a hen will start them up to three weeks old, and if they are started for about a week under a hen they may then be transferred to an incubator. We have never succeeded in hatching any egg *over* three weeks old at time of setting. If eggs are to be kept for a time before being set, precautions should be taken to prevent evaporation. They may be packed among green grass : our method is to wrap each egg in paper, place them upright in a tin box, and put a piece of wet blotting-paper in on the top before closing the lid. When we have hatched a sitting all from one hen the eggs have always hatched in succession, beginning with that last laid, and each day of difference in the age of the egg seemed to make a difference of nearly an hour and a half in hatching.

This difference would likely not be so great if we were not interfering with the hen or incubator during the hatching-out.

When setting such eggs we give the hen first those that are over a fortnight old : twelve hours later we give her those over a week old, and after another twelve hours the remainder. This causes all the eggs to hatch about one time.

When chicks hatch out before the expiry of the normal period of twenty-one days, it is sometimes taken to be a sign of strength : it is rather a sign that the eggs have been forced in some way, and the result is more likely to be ultimate weakness. The normal time is best, but we had rather have them a little late than too early. We are referring to the time of the hatch as a whole : if, however, one or two chickens hatch out ahead of the rest, that indicates strength in the individual, and ones that are much behind the others will generally be found to be weakly.

The nature of the grit influences, but does not control, the colour of the egg. Similarly you can influence the number and size of the eggs by the feeding, but in each case the control lies in heredity.

It is not expedient in a book like this to attempt detailed explanations of the principles which underlie breeding for a particular end, but some brief indication of important points may well find a place, since good mating is a vital matter for the breed—or any breed. We do not claim to have made a close study of these principles, but have read much that has been written on the subject, comparing it with our own experience. Theories are constantly being modified as research proceeds, and the whole subject fairly bristles with debatable points : it is difficult to understand : if you try to express it clearly and briefly you are forced into making general statements, although you are quite aware that numerous exceptions will be met with. But these difficulties need not make us avoid altogether this important subject, and it may here be said that though the following remarks are made in reference to breeding for utility points, many of them apply with equal force to exhibition breeding.

" Look well to the cock's dam." That is the general principle put into few words. Until you grasp the greater importance of the male bird your trouble will mostly be wasted.

The theory which best explains the phenomena, and so must be accepted in default of a better, is something like this. The male transmits his inherited characteristics—fecundity, size of egg, colour of egg, etc.—in some degree to his progeny of both sexes, but chiefly to his daughters. The female's influence is transmitted less widely : she influences strongly only her male progeny, and only a portion of these. Speaking generally, her influence on her daughters is remarkably slight. For example (leaving out of account occasional " sports "), when a male of the highest fecund type is mated to a female of the highest type, all the daughters are of this type, but only about half of the sons. The worst is that there is no certainty which of the sons are the useful ones till *their* daughters prove it by actual performance. Mr. Hanson's test offers some indication, and a study of his book, " The Call of the Hen," will be necessary to a correct understanding of it.

The subject of the inheritance of characteristics at once raises the question of in-breeding. The very word will be anathema to some readers, but the fact remains that in-breeding is the most effective method known of fixing desirable characteristics : it is also the most effective method of fixing undesirable ones. It is widely practised by the best breeders of utility fowls, and with very satisfactory results, but if it is not to lead to certain disaster it must be done by an intelligent breeder, and only with the best of birds.

Those who keep constantly crossing strains, and insist on mating birds that are totally unrelated, are likely to achieve neither disaster nor distinction. In-breeding leads to extremes—good or bad : out-crossing makes for mediocrity. When two unrelated birds are mated, considerable variation is the result, and as the progeny tend to resemble the nearest common ancestor—in this case far back—there is reversion to a much earlier type, with corresponding loss of what has been bred in since. If two different strains of high fecundity are crossed the average is lowered ; if two low-fecund strains, raised. The raising, however, is more marked than the lowering, since the added stamina from the crossing has the effect of increasing fecundity to some extent. The transmissibility of increased fecundity which is due to stamina is another question. Briefly, crossing, whether of breeds or of strains, improves stamina, but destroys the influence of ancestry. This can be regained by subsequent in-breeding or line breeding. It should be noted, however, that although stamina is often lost by in-breeding, the fault lies with the breeder, not in the system.

CHAMPION COCKEREL, 1913.
Winner of many prizes.
The property of Mr. H. P. MULLENS.

An in-bred male bird has higher prepotency, by which we mean the power of stamping his inherited characteristics on his progeny.

There would be fewer disappointments if both direct in-breeding and out-crossing of strains were left to men who have the requisite knowledge, the requisite birds, and the requisite patience to wait for results to appear in the second following generation. The sure and readily available way is to mate female crosses between two strains to a male bird—which need not be directly related—of one of the two strains, or the two-strain male to one-strain females. In other words, try to ensure that the nearest common ancestor is not too far back, and is likely to have been a superior bird. Before leaving the subject of in-breeding we would remark that analogy with the in-breeding of cattle fails in respect that in the case of fowls we have to deal with more prolific animals, and consequently have a wider range for selection. That stamina can be improved by judicious selection and in-breeding will not be doubted by those who know something of the development of pit game birds.

There are two stages in breeding. In the first you mate your best females to a male of superior quality, which you acquire from a good breeder ; later you come to a stage when you find difficulty in procuring a male bird which you are sure is better than your own, and as the male is the more important you then breed your own males, and take in fresh blood on the female side. This has the great advantage that you are able to test the bird's laying powers before using her, or at least before using her progeny.

Sometimes it is desired to identify the chickens of a particular hen. An incubator drawer divided by high partitions may be used, but it can be done under hens by enclosing the egg, when just about to hatch, in a muslin bag a little larger than the egg.

Culling is a matter on which everything turns if one is to breed good stock, more especially if one does not use trap-nests, and so cannot distinguish the eggs of the inferior birds with certainty. It must be practised early and often, and with much hardening of heart. The killing of a day-old chick is not a formidable or gruesome affair if you take it by the legs and swing the back of its head sharply against any hard object. We once saw a neat and expeditious way of killing adult fowls which may be of interest. A clergyman at a church sale of work, overhearing some one remark that he would have bought a dead bird, but that there were only live ones for sale, volunteered to kill one. A small crowd gathered immediately, but the bird was dead before they realised it.   Holding the bird on his left arm as one carries a bird, he closed his right hand on its head.   Letting go the body, he seemed to give the hand that held the head a slight backward jerk as the weight came on it, like the motion in cracking a whip, and the neck was dislocated.   He writes : " My manner of killing fowls was one which I learned from an old Orcadian sailor.   I was outside one day trying to kill a cockerel by drawing its neck, but I was not making a good job of it.   The old man passing at the time, I asked him how *he* killed a fowl.   Taking it from me he grasped it by the head with one hand, let it hang by the head, and gave his hand a circular motion until the neck was dislocated.   I always killed them in that way afterwards, and found it very effectual.   The older the fowl the greater its weight, and that aids the dislocation of the neck.   With a pullet or cockerel the neck is more easily dislocated.   I did not find it necessary even to tie the feet together, for in their kicking the toes do not reach so high as the hand that is grasping the head.   It is done very quickly."

As a saving of trouble it is convenient to have a distinctive colour of ring which you use solely for the purpose of marking birds that you wish to dispose of by killing or otherwise.   As soon as you decide that you will not use a certain bird for breeding purposes, you add this ring.   Croads, by the way, require rings of very large size—$\frac{3}{4}$" inside measurement for hens.   If you cannot procure rings of the size (usually numbered 9 and 10), small spiral rings may be expanded by pushing them on to a stick of the desired diameter and steaming them till they begin to uncurl.   Press the ends back against the stick and allow to cool for a few moments, when they will be found to retain the new size.   Rings of too small size should never be used.

When you feel disappointed with the birds you have bred it is well to remember that, as a result of growing experience, you yourself are becoming more critical : faults which earlier would have escaped you are noticed.   Under these circumstances you may feel tempted to change to a different breed, which may, indeed, please you better till you get to have experience of them, when, of course, you may repeat the process.   Your breed must suit your requirements and conditions, but if what you need is a dual-purpose fowl for heavy soil, damp district, or exposed situation, you are not likely to find anything better than a good strain of Croad Langshans.

A GOOD PATTERN OF TRAP-NEST.

The nest shown in the illustration is one of an American pattern of which a sketch appeared in *The Feathered World* some years ago. The important feature of it is the catch. It is made from a piece of wood 6 ins. by 2 ins. by about ¾ in. to 1 in. thick. The pivot is 2¼ ins. from one end, and the notch which holds the door is 2 ins. from the other. The face from *A* to *B* might as well be straight, or hollowed out a little if necessary.

When the nest is " set," the catch should be *more nearly upright* than in the diagram, and the door rests against the face *B C*. The hen, on entering, lifts the door with her back, and the catch falls back of its own weight into the dotted position, where the door should just clear it. The catch must be so loose on the pin as to work quite freely. If the wood of the nest is too thin to hold a nail firmly, a still better pivot can be made from a ¼ in. bolt with a nut on either side of the wood, the square part behind the head of the bolt being rounded off with a file.

D is a piece of wood of any convenient size, nailed on in a position found suitable by experiment, merely to keep the catch from turning too far in either direction.

Our nests are 20¼ ins. high, 22½ ins. from back to front, and 13 ins. wide—all inside measurements. The pivot of the catch is 6½ ins. from the floor, and the same distance from the front of the nest, but might with advantage be a little higher up and farther back : the best position will be found by trial.

A piece of strong wire with the end bent round is inserted in the front of the partition to check the door from swinging outwards. This can be turned up or down when the hen is to be released. The doors, of 3-ply wood, are hung from a piece of fence wire threaded at one end and tightened by a nut, but any system of hinges would do. The hinges are 4¾ ins. from the top, and the nest could be made lower by that amount if desired   The door must come almost to the lower front board, otherwise the hen will get her head under it. The hooked-on board is 5 ins. wide.

The door is shown in three positions : (1) Closed, (2) open outwards, (3) set ready for hen to enter.

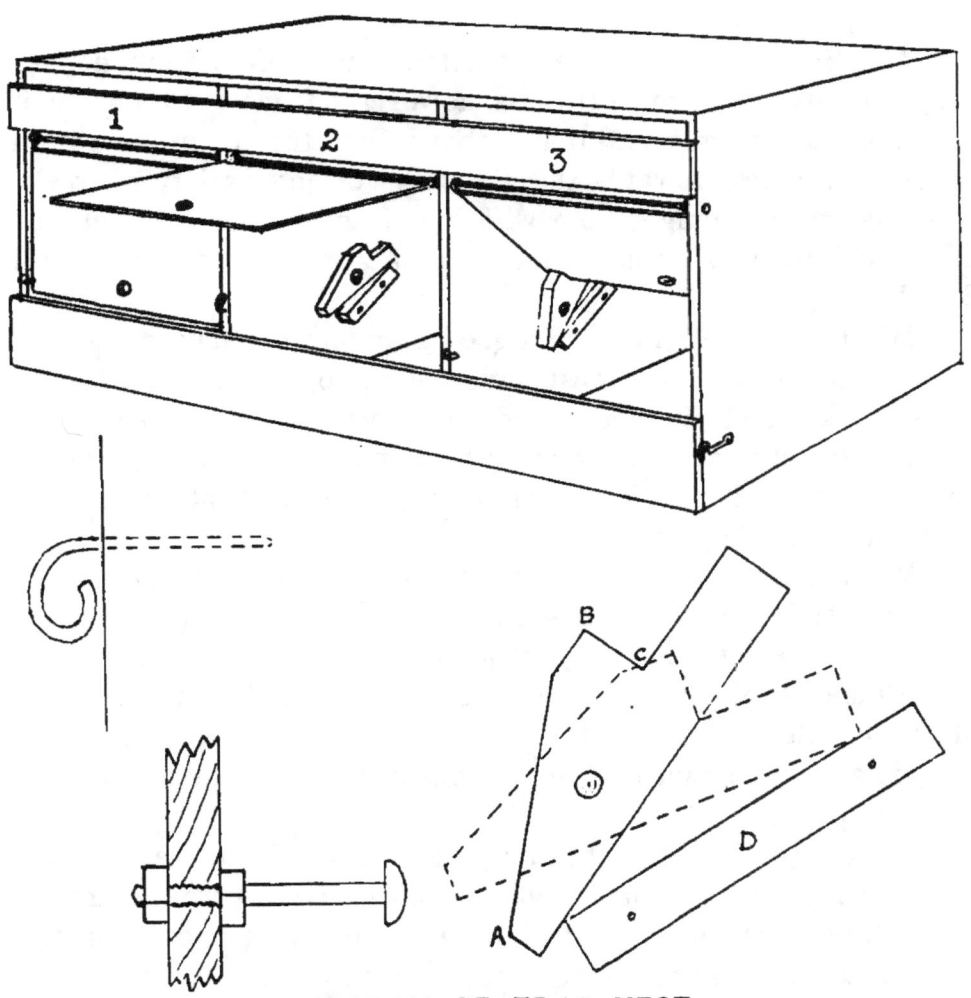

DIAGRAM OF TRAP NEST.
(See page 44).

PART II.

THE STANDARD.

The Standard of a breed is often—more often than not—a subject of disagreement. There is, however, one view of it on which all will cordially agree, and that is that among the conditions which make for progress towards the perfecting of any variety of fowls, and tend to encourage those who have not kept the variety to adopt and retain it, none is more important than having a good Standard.

What are the essentials of a good Standard ? In proceeding to answer this question we must have regard to two things. If you have to select a spade, you consider whether it is to be used by his majesty to plant one tree, by a forester to plant many, by a gardener for various uses, by a child on the sands. It is good only in relation to its purpose and user.

We must try to avoid having one Standard for the fancier and another for the utility breeder. The fancier will be helping the breed when he consents to slight modifications of his ideal to meet well-grounded objections from the utility standpoint, and the utility breeder will serve the same purpose when he takes into consideration the fancier's list of disqualifications in mating his breeding-pens.

Let us ask, then, for what purpose, and by whom, the Standard will be used. The Standard will, of course, be a word-picture, or word-photograph, of our ideal of a Croad Langshan. For the experienced breeder and judge the picture type of Standard is the better, concentrating attention on the most characteristic features ; his knowledge will serve for the minor details. But the beginner, if left to fill up the gaps from his own imagination, is little likely to fill them in correctly, and is liable to be discouraged when he finds, at some expense, that in some respect he has mistaken what is wanted. What is of more use to him is something more of the nature of a photograph, showing details. The danger here is that the attention is apt to be concentrated on the details, and drawn away from the broad essentials. Still, such a Standard is useful to those who, without buying experience at shows, yet serve the breed by propagating and spreading it.   These can do the breed much good or bring it into bad repute, according as they spread good or bad stock.   Any help given to such breeders in the way of making clear what is wanted helps to lessen the chance that a bird finding its way into a good yard will throw bad stock—a source of trouble with which breeders are only too familiar.

Our Club Standard is peculiar.   Unlike most breed clubs we have for our main object, so far as the Club is concerned, the preservation of purity of race, and only in the second place the perfecting of a certain type.   This accounts for the apparently undue importance attached in the Standard, and by specialist judges, to points which are of little or no weight in the case of other breeds. This also is the reason why hardly any but a specialist judge can judge Croad Langshans at all well.   A Croad judge, noticing any appreciable amount of black in toe-nails, a " full " eye, or a large head, sees in these things—small in themselves—indications of blood which is not Langshan.   We are not here discussing whether he is right or wrong: he holds that belief, and the bird is passed over.

Utility qualities should be kept well in view in the framing of a Standard.   Our Standard is practically the original Langshan Standard of 1877, with the wording of certain parts rather amplified, and one cannot help admiring the way in which Mr. Gedney inserted points of utility value.   His eye sees that for which he is looking, and at every turn in his description of his ideal we see that utility points are prominently before his mind.   Subsequent modifications of the original Standard, which have resulted in the revised Standard of 1913, have been mostly made by Mr. R. F. Housman, at whose hands the utility points have certainly not suffered.   In the correspondence which passed among the Committee of the Club regarding the revisal of the Standard and the relative importance the members attached to the various points, it was noticeable that points having any bearing on utility were keenly and critically scrutinised.

In 1919, when we happened to be in correspondence with Mr. Tom Newman, whose interest in, and knowledge of, utility breeding are well known, it occurred to us to send him our revised Standard with the request that he would criticise it from the purely utility point of view.   This is what he said :

" I have been carefully through your Standard, and certainly think that you have avoided extremes, and the Exhibition Croad should also prove a good utility bird.

" There seems to be no attempt to emphasise any particular point at the expense of others. Even with size you point to the necessity of fine bone, and I notice you do not penalise a bird for a few off-coloured feathers. . . .

" I am glad you held so strongly towards utility when revising the Standard, and also for preserving the one-pen character.

" I should like to see more points allotted to shape, which I hold to be the most important of all."

There has not come under the writer's notice any attempt to formulate a general Standard for " dual-purpose " utility fowls, but there are several such for the " laying " type, and in the absence of the former, the latter may teach us something. Let us bear in mind, however, that the type described is not the " general purposes " type, to which the Langshan belongs, and that we have in our own Standard a pretty accurate description of a " general purposes " type which has proved its utility—one from which we would not lightly depart. Still, none of us breed perfect fowls, and some consideration of the laying type " Standard hen " may lead to our erring on the safe side in our involuntary deviations from the Croad Langshan Standard.

Mr. Oscar Smart, in his " The Inheritance of Fecundity in Fowls," gives eight points as characteristic of a superior layer, insisting that the combination of these points, and not the possession of some of them, is the important thing.

1. Fine comb, long snaky head, and narrow skull.
2. A very red and prominent eye : the more the eye " stands out " from the head the better.
3. A long body, sloping very gradually towards the tail ; the tail itself being carried almost, but not quite, erect.
4. The breast-bone very short, and the abdomen (which should be covered with the softest possible down) well developed.
5. Legs rather above than below the average in length (but not too long) ; very fine and extremely pale in colour. The legs should be set wide apart.
6. Toe-nails extremely short.
7. Pelvis bones wide apart, not less than $2\frac{1}{2}$ inches.
8. Cartilage soft.

Mr. Tom Newman's ideas on what characteristics are found in a good layer are taken from an article of his on " How to Select the

Best Layers without the Trap-nest," published in the S.P.B.A. Annual Register of 1919.

"Our first thought must be stamina. . . .

"All coarse-looking and undersized birds should be rejected.

"The good layer is generally of medium size. The eye is bold and prominent. The comb is fine in texture, of medium size; avoid a "beefy" comb. The neck short and rather thin. Beak short, legs short, toe-nails short, breast bone short—the shorter the better.

"In a poor layer you will not get the width of two fingers between the pelvic bones and the breast bone.

"In a really first-class layer you will get the width of four fingers between the pelvic bones and the breast bone.

"A good layer is wide across the cushion and between the legs. The pelvic bones should be straight, thin and well apart. The back should be long and wide across the wings, the tail carried high and the bird tight in feather.

"Most good layers are late moulters and big feeders."

(Some other remarks are added which apply only to yellow-skinned breeds.)

In the same volume Mr. Joe Edmondson describes the "Standard Hen" as having a 9 in. length of back, a 5 in. width of back, and a 4 in. length of breast-bone, *or a proportionate measurement*. He goes on to describe the type he favours :

"A long and broad back, wide cushion, the legs well set apart and not too long, short breast-bone with a good distance between the breast-bone and pelvic bones, the pelvic bones should be straight and thin, and very pliable, sharp, bright eye, the head not too narrow, nor too small, neither must it be coarse, a strong, short beak, the comb of fine texture and not too large, the tail carried fairly high, the texture of the bones must be fine and the size of bird medium—4 lbs. for light breeds and 5 lbs. for the heavier breeds."

On first reading the foregoing the reader will most likely be struck with the differences from our Standard, but when they are studied closely, and allowance is made for the fact that they describe the "laying" type, and for differences due to what we may call special breed characteristics, there remains little that is not found in the Croad Langshan Standard, or, for that matter, in the Langshan Standard of 1877.

Looking at our Standard in detail, we find that it embodies most of those features which characterise the laying type of fowls.

While our main concern is to examine the Standard as bearing on the utility qualities of the ideal Croad Langshan, we propose, in touching on the different points of the Standard, to use some freedom in commenting on points which are not strictly utility points, but which may be of interest in a book like this. We happened to have correspondence over all the details of the Standard with a number of the Club judges, and retained most of the letters, recognising that these men had been at some pains to express exactly in writing what their ideas were, and that much was contained in these letters which would interest Croad fanciers. However, on going over the letters now with the intention of giving the benefit of this to the readers of the textbook, we came to the conclusion that this is not so feasible as it appeared to be. Most of the points of agreement are embodied in the Standard, and where there were interesting side-lights, the actual words of these breeders often depended for their interpretation a " universe of discourse "—a previous knowledge of their application of words— and taken out of their setting, expressed, or more often implied, they would convey meanings not intended by the writers. Some of these points may well be touched upon, but while in general we think it best to refrain from quoting our authorities, it is to be understood that we are not merely expressing our own ideas, and that our conception of what is wanted is largely derived from others, not least from Mr. Housman, whose highly instructive letters form a very large proportion of the aforesaid correspondence.

To follow the accepted order, let us begin with Shape and Carriage. An expression frequently used by breeders, which finds no place in the Standard itself, is " boat-shaped." What we want is " a large boat-shaped bird, fairly well up, but not too high, on leg." Now, this shape is common enough, but the difficulty is to get it in large specimens. Think of a Viking ship, with high, upright stem and high stern, well-rounded and rather bluff bows, body not tapering but of fair length, with good width throughout. A boat-shaped bird with good depth to the keel can show both plenty of breast-meat and good abdominal " capacity."

A difference of opinion will be noted in regard to length of leg of good laying type. It is a breed characteristic of the Langshan

CHAMPION PULLET, 1920.
Bred by Mr. A. BIRTWISLE.
(See Advert.)

to be fairly high on leg, and no alteration in this respect would seem to be desirable. " Shanks standing well apart " makes for laying, but when we come to " long breast-bone " we find that considerations of laying do not favour this. If, however, we get our birds with considerable depth, which is also required by the Standard, then the laying-expert will have little objection to a rather longer sternum than he likes. He will not object to our birds carrying plenty of breast-meat, if it goes with plenty of room between sternum and pelvic girdle, and if the bones of the latter are thin and straight. With a back as in the Standard no fault can be found. The open fan-shaped tail we like to see is certainly not against our birds, giving, as it does, a wide appearance to the stern. The " tail carried rather high " is also a good utility point : the angle, by the way, is about 75 degrees, according to Mr. Housman. The wide, fan-shaped tail causes a curious lyre-like effect of the sickles in a cockerel possessing a good tail, not in a cock.

The whole section on Size, Bone, Flesh and Skin is directly in line with utility. The only part of it which might be challenged is the prominence given to size, but there is the qualification, " consistent with type." We put a query lately to a friend, who, as the saying is, has forgotten more than we ever knew about breeding—both for exhibition and utility. He never bred Croads. After answering the question, he added :

" If you are revising Croad Langshan Standards, I think you should fix for size, and deduct for over Standard size as well as under, i.e., if you are to breed for eggs. Of course, if to be bred only as a flesh-former, then the larger the better. The importance of eggs is such that the overgrown should not be encouraged. Is this correct ? "

Personally, we should not care to *deduct* for over Standard size, but undoubtedly our friend is right in laying stress on the fact that eggs are the more important output, and we venture to suggest to our judges that size beyond what is reasonable in so large a breed should not be allowed to weigh in a bird's favour, since such a bird is not likely to breed good layers. Mr. Oscar Smart, however, in the work already mentioned, has this :

" But it has yet to be proved that a big bird is necessarily an inactive one. That some are, we may allow, but this appears to us to be more a question of strain and of breed than of size. Thus we have known Light Sussex, Rhode Island Reds, and

Croad Langshans quite as big as some Black Orpingtons, but they have been infinitely more active and much better layers. There appears to be a good deal of evidence to prove that lethargic birds are poor layers, but very little to prove that size causes lethargy."

Under " Plumage " we have little comment to make. Anything in the way of useless ornament is regarded as being against laying. Leg-feather is a breed characteristic, but much leg-feather is undesirable, and heavy leg- and toe-feather, such as we often see, is not desired by the Standard. " Feather soft, neither loose nor tight " is quite right, since the Croad, being a bird suited to a severe climate, must not be too tight in feather if it is to lay in winter. The " full " tail, not advantageous from a laying viewpoint, is perhaps a natural concomitant of ample breast muscle and consequent wing-power.

In the last section of the Standard, that on Head and Feet, we find more points worthy of notice. The head is small : it is again a breed characteristic, and we are not out to manufacture a new laying breed. " Full over the eye " has been much debated. If we mistake not, it was taken out of the Standard, and later was re-inserted by Mr. Housman. A heavy-browed bird will cause the utility men to lower their brows. It is worth noting that the expression is used in the earliest Langshan Standard, so presumably a number of the first-imported birds were " full over the eye." The comb as described is satisfactory, and no change should be made in either direction as to size of comb. Too large a comb, besides spoiling the bird's appearance, is against laying in severe weather conditions : it reduces the bird's temperature too greatly in cold weather. On the other hand, a very small comb may be thought neat, but it allows the temperature to rise too high in hot weather, inducing excessive broodiness. The scales on the shanks are preferred to be nearly black in young birds, but this point is not so important for exhibition purposes as it is in most black breeds. Let us say generally in regard to this section on head and feet, that a good eye and comb are of importance for both utility and exhibition, but that most of the other points (toe-nails excepted) in the section are not, in the estimation of judges and breeders, nearly so important as they seem to be in the mind of the novice exhibitor. We feel sure that many a bird that could be in the prize-list at some of the " big " Shows is kept at home on account of mid-toe feather or some such defect—we had almost said " trifling " defect. If you have a bird with shape, size and colour enough to catch the judge's eye, a few minor faults will not pull it down. A point which is not stated in the Standard is this, that a Langshan should not be " full *in* the eye " ; the white of the eye should be frequently visible as the bird looks about ; more particularly is this noticeable in a male bird.

Last in position in this section, but of first magnitude, comes the thorny subject of " Black in toe-nails." If not in itself a utility question, yet, as vitally touching the popularity of the breed, it is one in relation to which we all want to know exactly where we stand. It is made fairly clear in the Standard, but the reasons for it are not. The Langshan Standard from the first said " Toe-nails white." The writer has not been able to ascertain whether the Langshans in China had the toe-nails entirely white or not ; probably at the source they had, or nearly all of them had. It is just possible, however, that until attention came to be focussed on this, later, some marks may have escaped notice. Let us suppose, however, that all were pure white. It is now considered that black in toe-nails is an indication of cross blood somewhere in the ancestry. Whether this be true of not, we may remark in this connection that in a human being a trace of Negro blood shows in the nails where there may be no other indication. It seems reasonable, there-fore, to suppose that the nails may afford a sensitive test for certain crosses in fowls. The position now is that a great proportion of Croads, even in the best yards, come with some dark markings in toe-nails. The best breeders, often at no little loss, reject such birds, and endeavour, more or less successfully, to breed only birds with clean toe-nails. The trouble is wide-spread, if not everywhere. By careful breeding and rigorous selection breeders have succeeded in breeding clear of this trouble, but then their difficulty is to get fresh blood without re-introducing it. That a bird is itself clean in nails is no guarantee that even with a mate also clean it will throw progeny all clean in nails. If you breed clear of it, you may by mating different lines of your own continue to get white toe-nails, and yet, all the time, the tendency to throw this fault may be latent in your birds, only awaiting the opportunity afforded by a sufficiently distant mating to manifest itself. You take in a cockerel for fresh blood, mate him to your white-nailed hens, and some chickens come with black. The cockerel gets the blame, of course, since you have bred only white for some time back. But the matter is not so simple as that, and it is by no means certain that the cockerel is to blame, or that he alone is to blame, though he is let in for the role of scapegoat. If it could be proved that the defect is capable of being " bred out," then there could be no objection to the heaviest penalising, but this has yet to be proved. In the revising of the Standard the general consensus of opinion was found to be in favour of making the existing exhibition requirements clear by making " other than white toe-nails " a *disqualification*, instead of a *defect*, as it had been formerly. It was felt, however, by the majority of the committee, that this change was more than was warranted, and as a compromise it was agreed to transfer the expression to the list of disqualifications, but to modify it so as to exempt from total disqualification a bird which showed no more than a mere trace. There was a great deal of discussion, two of the members whose word has great weight expressing the opinion that to allow any black at all was " the thin end of the wedge." It was quite clear that every member of committee was anxious to take measures to reduce, and ultimately stamp out this defect, the only difference of opinion being as to whether extreme stringency in the Standard would have the desired effect, some thinking it would, others that it would do harm by making the breed less popular. After all, the question resolved itself into this, and thus it was put to Mr. Housman, who favoured extreme measures. " Would you ' place ' (i.e., give a prize to) a bird with a mere trace of black in nails ? " His answer was that he would be prepared to do so if the bird, in his opinion, had sufficient superiority in other respects. That cleared up the position which it was desired to elucidate. The whole subject of breeding is difficult. No amount of knowledge will keep one clear of faults. For example, a bird clean in the toe-nails, which may have been a " big winner," will sometimes discover this fault later in life, say, when three or four years old. By this time it has been bred from, and its progeny widely used.

Our apology for mixing up utility with other considerations is simply this : We wish every utility breeder to do what he can to help the fancier to preserve the character of the breed, and every fancier be an enlightened utility breeder. Mutual help is not possible in the case of some breeds : with the Croad Langshans it is possible, and the Club is looking for it.

PART III.

SOME EXTRACTS FROM MISS CROAD'S BOOK.

The name " Croad " Langshan, used to distinguish the lower or " Dorking " type from the taller " Modern " or " Club type " Langshan, is a tribute to the lady who fought hard, and in the end, successfully, against the Black Cochin breeders, who eagerly claimed the new arrivals as Black Cochins, and did everything in their power to prevent their being recognised as a new and distinct breed of fowls.

Miss A. C. Croad, of Durrington, Sussex, was the niece of Major Croad, of Durrington, who introduced the Langshan into England. The association of her name with the lower type of bird commemorates her preference, and that of her uncle, for the type which now bears their name. It was, however, only a short time before her death, and after the death of Major Croad, that the Langshan camp became divided over the question of the two types.

In furtherance of her campaign to secure recognition of the Langshan as a pure and distinct breed of fowls, Miss Croad published a book entitled " The Langshan Fowl : Its History and Character- istics." This work has long been out of print. For a copy of the third edition (London : Bowers Brothers, 89, Blackfriars Road, 1889), from which these extracts are taken, we are indebted to Mr. R. Fletcher Housman. The copy itself is of special interest. A letter from Mr. Frederick Geeson inside it begins thus : " I am sending you the enclosed book. Some twenty or so years ago the book was sent me by Miss Croad, . . . and I think she said the book was her last copy. The marginal marks are hers." The marginal marks are numerous, and indicate those portions to which Miss Croad herself attached most significance.

The book extends to 122 pages, with an Introduction by Mr. C. W. Gedney. This gentleman sacrificed his own interests and came to the help of Miss Croad in the early years of the struggle, when he realised that the forces arrayed against her, led by Mr. Lewis Wright, were bearing her backwards. Mr. Gedney it was who drew up the first Standard of Excellence of the Langshan, in 1877.

Soon the tide turned, and in 1884 the Langshan Society was formed, with Mr. Lambert as President.   Miss Croad was a member of Committee.

The first part of the book consists of a most able presentation of the case for the recognition of the Langshan as a pure and well-established breed, and a merciless examination of the points of the Black Cochin and the arguments of its champions.   The frontis-piece is a reduced reproduction, measuring 13½ by 9 inches, of an Admiralty Chart of the Yang-tse-kiang from Shang-hai to Nan-king.   On this one can make out, about 100 miles above Shang-hai, the names "Lang-shan Crossing," "Lang-shan Flats," and a little to the north of these "Lang-shan Pagoda."

Now let us quote from the book itself, omitting the controversy, except for a few references to those who helped therein.

A letter bearing date November 27th, 1871, conveyed to the late Major Croad intelligence that a nephew residing in the north of China had made the purchase of a new breed of fowls for him. Successive letters made mention of these birds :

" The fowls I am sending you are very fine ; their plumage is of a bright glossy black.   I am told that their flesh is excellent. The Chinese say they are allied to the Wild Turkey.   They are valuable birds—you must be careful of them, and get them acclimatised by degrees."

. . . When the fowls made their appearance here on the 14th of February, 1872, they looked somewhat jaded, and were at first very shy. . . . There had, however, been no casualties.   They soon grew familiar, and in less than a week their combs and wattles resumed their brilliant red, and their feathers their glorious sheen. On the 16th (two days after their arrival) the hens commenced to lay, and we then made a shrewd guess, which has since been amply verified, that these splendid creatures would prove as hardy as they were beautiful, and that no gradual acclimatisation would be needed.

. . . In the autumn of 1875 we received a letter from Mr. Gedney, addressed to the late Major Croad, inquiring if we still kept our breed of Langshans pure, etc.   To this we replied, giving the history of our Langshan troubles, and the cause of our birds' retirement from public life ; and from that date he has worked with us in our endeavour to obtain that recognition for the Langshan to which it is fully entitled.   Through his exertions a separate class was obtained for them at the Bromley Show in 1875.

So powerful had the opposition proved, that although we had occasionally advertised Langshan eggs, and birds, in the Poultry journals and local papers, we only sold *two sittings of eggs and three birds* from 1872 to 1876. . . .

Mr. C. W. Gedney . . . determined not to sell a bird or egg in this country until the breed had become thoroughly established, lest his advocacy should be laid to the charge of " interested motives."   . . . and he it was who shipped the first Langshans to America, to a Mr. Samuels, . . . but, like the Langshan in England, they met with most determined opposition from a section of the " Fancy." . . .

In the autumn of 1883 a fortunate circumstance brought the Langshan to the special notice of Mr. Harrison Weir. . . . When he understood the nature of the conflict he at once decided to draw a lance in the services of the breed.   Mr. Weir's testimony is of high value, for he has from his earliest days been an ardent ornithologist and fancier, besides being an animal painter of no mean repute.   The children of this and succeeding generations will have good cause to remember the genial artist. . . .

(Doctor Gabb, of Bewdly, also brought his ornithological knowledge to the help of Miss Croad.   Mention is made of Mr. Bennett, of Elgin, and Mr. Lewis P. Muirhead, Helensburgh, the latter of whom got eggs in 1876, and in 1882 published the results of a most interesting experiment in crossing.)

. . . Mr. H. Morrell, of Headington Hill Hall, Oxford, . . . originated the classes for Langshans at the Bath and West of England Show, and . . . for many years sustained them *by giving all the prizes*, this . . . when its very name was only a signal for contempt and abuse.   Besides . . . Mr. Morrell has given un-numbered cups and prizes. . . .

(M. V. La Perre de Roo, France, and Captain Sargant and Mr. Kesling, America, are mentioned—the last-named as having had three White birds among two lots imported by him, which he describes.)

⊢ One gentleman . . . told us he had been in the Imperial service of China ever since 1859, that he had travelled many thousands of miles in the interior in all directions, and had never come across the Langshan in any other part except by importation . . . and hat *Lang* should be translated " two," and *Shan* " hills."   He

"HIS MAJESTY."
Winner of many prizes for Mr. R. O. RIDLEY.

added that he and other residents in North China well remembered the introduction of the Langshan to the European community; its date was fixed by the placing of a Light-ship outside the Langshan crossings in 1862. The officers and crew of the Light-ship landing to explore and forage, came across this fine breed of fowls, and as occasion offered would send presents of eggs and birds to their friends in Shanghai. . . .

There can be no doubt that the Langshan is a bird whose origin dates very far back. When asked how they came there, the Chinese invariably reply they " do not *know* ; they were *always* there," and the bird itself bears out this theory entirely, for, as Dr. Gabb remarked, " It is one of the purest and most distinct of breeds." The priests at the temple look upon them as " Joss," or sacred birds. At one time we know they made great difficulty in parting with any specimen. When they are in full moult is the outer " barbarians' " best time, for they are then considered unfit to offer to their gods. As the Chinese follow the heathen custom of feasting on the meats offered in sacrifice, a Langshan in good condition would, no doubt, be a plea for themselves. . . .

The back of the nestling is covered with black down, and the head, face, and breast are a mixture of black, white, and different shades of canary colour. These shades are by no means distributed according to rule—in some the light predominate, and in others the dark. A chicken with reddish down, or one all black, should be rejected. The legs, when the chicks first emerge from the shell, are, in some, pinkish, like a young Dorking's, in others they have already assumed the dark pencilling peculiar to the older birds. Whilst losing their down there comes a time when they are not attractive-looking, but the appearance of the plumage during this stage depends greatly on circumstances—the shelter and care afforded them, the food with which they are supplied, etc. When they have donned their first black coat they have much the appearance of young turkeys, but it is not until they are about five months old that the cockerels and pullets give indications of their future grandeur ; every day then finds them increasing in size and beauty. They are admitted on all hands, by those who have kept them, both in England and France, to breed very true to feather, and this is the strongest test of purity. The young birds of both sexes often retain their white nest feathers until almost fully grown, and during this period occasionally have white or a tinge of white about them ; this oftenest occurs in the primaries of the wing. . . . Now and then a Langshan cockerel will show red or golden feathers, which generally appear in the hackle or saddle, one or both ; these coloured feathers always come with the first adult plumage ; they are repeated, feather for feather, after every moult, and one generally finds that these specimens have only taken another form of beauty, for we have occasionally seen a cockerel that looked as if a shower of gold had descended upon him—the yellow plumage, having all the lustre common to the black, shining in the sunlight like burnished gold.   The hens never contract coloured feathers, but, like all other black breeds, are liable to white.   As we before stated, they are not all cast in the same mould : some stand high, others are closer to the ground ; some have a tolerable amount of leg-feathering ; others, again, have little, and, in rare instances, are nearly bare-legged.   In some the combs droop, in others they are erect, and with the edges fine and evenly serrated.   Some have black eyes, others hazel ; and all these various types (if we may so call them) were represented in our imported stock, so we may look upon them as accidents and *non-essentials*, for in points of intrinsic merit all agree.   The same glorious brilliancy of plumage, the soft, yet vivacious, expression of the eye, the erect carriage, the neat tapering head and neck, arched hither and thither at the slightest sound ; the white and delicate flesh, the light framework—these are the heritage of all the family. . . .

We have been told there is an individuality in hens' eggs, and that the person used to collecting them can generally divine by which hen each particular egg has been laid.   Be that as it may, the Langshan seems to defy all set rules in this respect, and indulges in a charming variety ; the tints are varied from the palest salmon to the darkest chestnut brown.   On some there is a bloom like that on freshly-gathered fruit, whilst others are spotted, often literally splashed all over with dark spots, and the same hen will tint her eggs differently one day from what she does on another. We have noticed that these spotted eggs occur most frequently during the spring months, when the secreting organs are most active and the calcareous matter which forms the shell is more readily obtained by the hen.

Langshans are unrivalled as the layers of rich, medium-sized eggs ; many of the younger hens lay during a greater part of their moult, but this is an exhausting process, and should not be encouraged. . . . The pullets usually commence to lay at about five months, but the early matured birds do not invariably prove the finest; for our part we prefer their being less precocious. A pullet that begins to lay, say, in November, will often go through the winter with little interruption, and will, moreover, give her possessor an immense number of eggs before she evinces the least desire to sit. A breeder of these birds complained that up to May of last year (1876) not one of nine hens had asked for a nest. . . .

The Langshan is a remarkably thin-skinned bird, and this is especially noticeable in the red appearance down the leg and between the toes. For this reason a chicken of this kind is always considered difficult to pluck, and, indeed, it is so fragile that a touch of the knife will almost joint it when brought to the table. The legs are dark pencilled; in the older birds (especially during moult), the colour fading into a silver-grey; and the legs are, moreover, slender for the size of the bird, the toes being more flexible than we have noticed in any other large breed of fowls. . . .

The Langshans are very hardy and easily reared, but they cannot stand pampering; if you want to succeed with them, let the mother rear her brood in the open air, and, so long as they have a dry floor, and protection from rain, they will thrive in spite of frosts and east winds. We have reared January-hatched chickens in the open air without the loss of a single bird, whilst others, "coddled" in comfortable quarters, have sickened and died in the most exasperating manner. The adult Langshan will submit to confinement, and do well if enforced, but prefers liberty and a wide range. The cocks, if pursued, will scale a wall of considerable height; the hens, though active, are not so venture-some.

. . . All the Langshans we have ever seen (and these must number many thousands) carry their tails somewhat high. This has been described as "boat-shaped." . . .

. . . One of the cocks in our first importation had a yellow neck hackle—it was so brilliant we used to call him "golden throat." Two others came at different times with our earlier importations, and we bred from these. . . .

. . . We have always been careful to obtain a clean hock joint, and have avoided exuberant leg-feathering. . . .

. . . Mr. Comyns quite appreciated the true bearing of a Langshan when he said ·it should be more like a Game fowl than a Cochin ; but that is not saying the Langshan should be like a Game fowl. . . .

. . . It is not because the Langshan Club desire a bird as long on the leg as possible, that its members would for a moment suggest in selecting this bird (tall type) ; they do it from a desire to breed *away* from Cochin characteristics ; these latter the Langshan *never possessed*. The taller type of Langshan is one we prognosticated would have many admirers. In its hobble-de-hoy stage it is the more ungainly bird of the two, but when fully matured, a perfect specimen is elegance and grace personified ; it is unique in form, and unlike any other breed. . . .

. . . In 1878 Mr. R. Fletcher Housman commenced as a Langshan breeder, and he has always aimed at keeping his stock pure ; we do not mention him as being peculiar in this respect . . . but we believe the birds in Mr. R. F. Housman's yard led to one of the most valuable conquests the Langshan has ever yet made. Mr. William Housman had naturally many opportunities of observing the Langshans in his brother's yard . . . he decided on keeping Langshans himself . . . and they have inspired many of the most interesting poultry articles that have appeared in *Live Stock Journal*.

. . . The Langshan has also proved useful in the manufacture of other breeds of considerable value—we will only instance the Orpington. . . . Mr. Cook visited our yard in January, 1887, when he told us the more he saw of the Langshan the more valuable he had proved it to be ; he had at that time commenced mating his Orpingtons, and we supplied him with some birds as nearly bare of leg-feather as we could find for his purpose. Mr. Cook has been very open in his statements regarding the Orpington ; we know exactly how the breed has been made. When the Orpington Club was instituted they did us the honour to invite us to become President, an honour we were, however, obliged to decline. . . .

. . . We think against coarse skin and bone there should be a heavy mark in " deductions " : a mark that would amount to extermination. . . .

The account given by Major Croad of his birds was a minute and faithful one. (This refers to the original importation, and, in particular, apparently, to the birds he exhibited at the Crystal

Palace Show in 1872.) " The old birds were in full moult ; one of the cocks was of the tall, up-standing type, the other of what has been since called the Dorking type, with full tail ; we cannot remember, however, in what condition it appeared. Two of the hens had fine erect combs, were large birds, and symmetrical in shape ; the other two hens had slight tufts, these tufts we believe to have been mere 'sports.' One thing we, however, proved, we did not obtain any tufted chicken from the erect-combed hens, but from the tufted we did."

The foregoing quotations from Miss Croad's book have been chosen for descriptive or historical interest. It must be remembered that they refer to the Langshan, and not to one type only. The various editions of the book, differing slightly, were published in 1877, 1880, 1889 and 1898.

Another controversy began in the winter of 1902 between the supporters of the two types, culminating in 1904 in the secession from the Langshan Society of those who favoured the lower type, and the formation of the Croad Langshan Club. Mr. R. F. Housman was the prime mover in this matter. The Croad type of bird had ceased to have any chance of winning in Langshan classes, and declined in popularity in consequence. Happily it was cared for during these years, and at no small sacrifice, by a few whose names are to-day familiar to all Langshan fanciers. At length Mr. Housman made strenuous efforts to get the standard altered to admit the lower type on an equal footing. This met with equally strenuous opposition, as was to be expected. Even as in the ecclesiastical arena the most bitter strife is found where the tenets of the opposing sects differ least, so the supporters of the two types of this excellent breed dealt each other hard blows for the supremacy of their own favoured type. Each accused the other of crossing with alien breeds—Cochin, Game, Malay, Orpington and " Black Diamond," among others. The last mentioned was a term used to describe a Langshan-Cochin cross imported in good faith from China.

This internecine struggle seemed likely soon to leave the Langshan without any position worth fighting for, and the best way out of it was adopted by Mr. Housman, when, in 1903, he drew up a modified Standard, called the Pure Langshan Standard, and asked for support to form a Pure Langshan Club. Objection was taken to this name, and it was modified first to " Pure Croad Langshan," and then to " Croad Langshan," and the present Croad Langshan Club came into being in 1904. Both parties were then free to direct their energy into channels more profitable to the breed.

Quite recently proposals were made for a compromise such as Mr. Housman sought. The trend of the resulting correspondence showed a friendly spirit, free from the bitterness of by-gone days, but the general opinion seemed to be that the points of difference, such as length of leg, tight and soft feather, hard metallic green and softer yellow-green sheen, and fullness of tail, having now become more firmly fixed in the interval, would render inexpedient, if, indeed, not altogether impracticable, any attempt to amalgamate the two types.

Breeding for Exhibition.

By Herbert P. Mullens.

Mr. H. P. Mullens.

IN these post-war days poultry breeders and the poultry Press, and, to a certain extent, the daily Press are all talking of utility poultry until the general public, with the exception of those who have kept and bred poultry for many years, have formed the opinion that utility poultry and exhibition poultry bearing the same name are virtually two distinct breeds. The exhibitor has himself to thank for this in that he has so frequently neglected the utility qualities of his birds for the sake of show points ; and there is a good deal of ground for this opinion. If one goes to see a flock of utility White Wyandottes, for example, what does one find ? A flock of yellow-legged white birds, small and almost devoid of true Wyandotte type, layers of a small egg, and though a few in the flock may produce anywhere from 200 to 240 eggs in the year, the main body are laying about 170, or less.

I am often asked, and the question is often asked in the Press, " What is the best utility breed ? " There is no best utility breed. But what is the best utility fowl ? Is it the bird that can produce a very large number of eggs, the bulk of which are barely two ounces, and being a small bird is not the best for table qualities ? Or is it the bird that can, and does, produce 200 or more large eggs per annum, is, when fatted, an extra good table fowl, and can, and does,

"A STUDY IN TAILS."

Mr. H. P. Mullens's Cock.

(See page 68.)

win in the show pen, or is, at any rate, up to show form ?   There can be only one answer to that.

My readers may say, What has all this to do with the Croad Langshan for exhibition ?   My answer is, It has everything to do with exhibition ; it is, in fact, the foundation on which alone success as an exhibition fowl can come to any breed, and more especially to the Croad Langshan.   There is a dislike to a feathered-legged fowl, why, I do not know, but that dislike exists and Croad Langshan fanciers must do something to counteract that dislike. The way to do so is to pay great attention to the laying and table qualities, which by nature cannot be excelled by any known breed of poultry ; but if these qualities are neglected for merely show points the breed will quickly lose all interest in the eyes of the general public.   Therefore, let all Croad Langshan breeders remember this : *The two great foundation stones of breeding exhibition stock are laying qualities and table qualities ;* when you have these, see that you maintain them and on them build your perfect exhibition fowl.   It can and has been done, and if only breeders would combine to breed on these lines the day is not far distant when Croad Langshans will be one of the most popular breeds in the country.

Now as to breeding for show.   Taking for granted the utility points are there, and taking the cock first, select first of all a bird excelling in type.   The Standard and photographs in this book will explain the desired type.   He should be active, carrying his head and tail high, flat across the shoulders, back rather short, with a medium cushion rising rather abruptly to the tail : the latter should be very wide and full, so much so, that when the bird has his tail towards you it hides the bulk of his body.   I am sending a photo with these notes which will emphasise what I mean.   Select a deep-bodied bird, standing on medium length legs set well apart. As to size, the larger the better, provided he is active.   See that he is sound in his feet, and that they and the toe-nails are the correct colour.   Having all these good points in your male birds, select the best-coloured one to head your breeding pen, as I find that a good coloured male mated to even dull-coloured females will produce good-coloured stock, whereas I have never known a poor-coloured cock throw a good-coloured chicken.

As regards the hens to mate with such a cock, other things being equal, select the largest, as you will get your size from the hen. Having your pen mated, make a rule never to use a small egg for hatching, and if you are keen on a deep-coloured egg, select for colour as well.

The Croad Langshan is an easy bird to prepare for show. Personally, I never train my birds for the show pen until about ten days before the show, and I find that in most cases two days training in a show pen is sufficient. If you want your birds to be in good bloom do not coddle them and keep them shut in for days before showing. Let them have their liberty whatever the weather may be, until the day before they are sent to a show; by doing so you will secure nature's bloom. The judge, if he knows his job, will go first for the correct type, and under a specialist it is useless to show a bird that is not typical: great size, high colour, perfect condition—none of these can carry a bird to the top if type is missing. You must have the correct type, and, after nearly forty years' experience of the breed, I have no hesitation in saying that the better the type in a pullet the better the laying qualities, and a good layer is always in good condition, which is only repeating what I have said before, viz., utility qualities are the foundation on which alone the exhibition fowl can be built.

What is Laying Type?

By W. POWELL-OWEN, F.B.S.A.

(On Council of National Utility Poultry Society.)

MR. W. POWELL-OWEN.

WHEN requested by the Secretary of the " Croad " Club to write an article for this textbook, I gladly promised; and I could think of no more interesting or topical subject than " Laying Type." Helpful criticisms and discussions lead on to progress, which is my excuse for being critical, and for dealing with a controversial matter. I must say right away, however, that I am not a utilitarian extremist, but have always been a fancier with a leaning towards " utility with beauty "; my views of to-day coincide with those I held prior to 1914, and which, in those days, were often responsible for my being referred to as the " utilitarian-fancier."

The hen has always interested me as much from within as from without ! And long experience in post-mortem work should teach one what duties each organ performs and where the eggs come from if nothing more. What are responsible for egg-production ? is a query that should be of more than ordinary interest to all poultry-keepers. The following play an all-important part : (1) Breed ; (2) strain ; (3) rearing ; (4) feeding ; (5) skeleton, and

(6) selective breeding. Individuality is a remarkable factor in heavy laying, consequently calling for selective breeding and strain. Feeding the layers is also very important, because you can readily ruin, by injudicious feeding systems, birds that would otherwise lay well. Next we come to skeleton, and upon this peg I hang my cap as representing " the hen from within." Even sisters out of the same dam and by the same sire vary in production, and mainly because they vary in the make-up of the skeleton.

I have not yet handled a super-layer that was narrow across the back, which is quite natural seeing that the ovary, or un-developed yolks, are situated just there. I take my measurement across the back, from the top of the left thigh to that of the right. Maximum points in my utility score-card go to a bird measuring five inches or more across the back at this spot. My next measurement is between the legs, and in my super-bird that takes the full points I can get five fingers (about $4\frac{1}{2}$ inches) between the legs. This measurement should be taken while the bird is standing on a firm base, and not when raised off the ground or bench, or the bird will spread her legs and register false. I consider these two measurements as representing almost a third of the capacity of a laying hen, i.e., as denoting the eggs within her. It means the capacity of the ovary.

What has the Standard for Croads to offer me ? First of all, it is stipulated that the back shall be broad and flat across the shoulders. As regards leg-carriage it says that the shanks shall be well apart. Such items are sufficient to place eggs in plenty inside the Croad.

Nature places egg-production last, whereas we place it first. If we are to insist upon that order we must help the hen. The laying hen requires so much of all food given for maintenance, and a given quantity beyond to help her fight unfavourable elements that may come along, to repair wastage in bone, flesh and feather, to pump blood all over the system, to make up deterioration of laying organs . . . and any surplus goes to make eggs. Now, if a bird has abnormal headgear, blood has to be pumped into it, and the ovary and oviduct are robbed of it ; if a hen has excessive feathering, bone or flesh, it must repair all waste and deterioration before it can think of making eggs. A bird with surplus feather, etc., must also take longer to moult, which means another reduction in output.

Fine texture without abnormalities is a very important factor, and the Croad Standard votes for a medium comb of fine texture and one rather small in size.   That is most useful as it releases blood for egg-making.   Bone, too, must be " medium or rather fine," which is another utility merit.

A controversial subject rather centres around length of breast-bone, which, by the Standard, should be long.   In a laying hen abdominal capacity means a great deal, as a view from within will prove.   The hen, unfortunately, is built on the wrong lines. The oviduct should have been placed on the opposite side ; instead, it is directly above the gizzard and the intestines, and nature again gives the latter organs the preference.   When a hen is out of lay all the organs decrease in size, but as she comes into laying condition they increase and call for elbow room.   If you have a long breast-bone the latter acts as a stone wall ; consequently, as the organs increase and fill the abdominal capacity, the gizzard is pushed up against the oviduct and interferes with laying.

The majority of birds sent to me for post-mortem examination have long rounded bones and an absence of abdominal capacity. I like to get five fingers (maximum) between the end of the breast-bone and the pelvic bones in my super-layer.   Then, when the organs swell they can push out the skin and drop downwards instead of going upwards on to the oviduct.   For the same reason I like fine texture of skin as against any coarseness or "stone-wall" effect.

Anyone who has plucked and trussed a fowl will know the " yards " of internal organs he withdraws from the small abdominal capacity.   There is no room for even a little surplus fat, unless it be at the expense of laying.   When the fat lines the abdomen and starts attacking the gizzard and the other organs, the long-breast-boned bird soon " goes west."   Digestive disorders set in, or ovarian troubles, all because the organs are pushed up instead of going downwards or lengthwise.   As an escape or safety valve I like a shorter breast-bone, reasonable length of abdomen, with depth and width of abdomen also.

With the Croad, too, we are dealing with a table breed— one that has inherited the tendency to fatten readily.   One, too, that goes on fattening with age.   Hence the importance of abdo-minal capacity.   Many may suggest the long breast-bone for those breeding table chickens, and the short one for egg-production ; it is a subject worthy of careful thought and discussion.

Excessive size gets us further away from egg-production for the reasons of repair and deterioration already mentioned. My ideal of a Croad is a bird of medium size without too high a tail-carriage, or too long a breast-bone or rounded. The medium-sized bird is ever active, matures more quickly, and is the better layer. And I would like to see all breeders concentrate upon the deep-coloured egg, which has always been associated with the Croad ; it is a factor that must never be lost, and by setting selected eggs we can retain that rich colouring which no other breed can equal.

The Exhibition Standard—Some Discussional Points.

By W. POWELL-OWEN.

I HAVE been requested to criticise helpfully the Croad show type as far as utility goes. Many will disagree if I translate utility into " egg-production," but, surely, an ideal table breed like the Croad can look after its table properties and yet give us laying above the average. It has the natural tendency to put on flesh and to fatten ; consequently, the material is there even in the laying type for all to mould by special feeding if their objective be table produce. My ideal utility bird in such super table breeds as the Croad and the Sussex is the medium active bird that will lay the eggs ; we want eggs ere we can get our table chickens, so that breeding stock in such breeds should be good layers. But in dual breeds, like the Croad and Sussex, I am not an extremist and do not hanker after high individual laying such as might come from the Wyandotte or the Leghorn. My ideal is heavy laying in so far as general plodding throughout the year, and especially in the winter months, ending up with yearly totals well above the average.

In a layer I do not like the extremes in legs—too high or too low—but sufficient length to show the hocks as the Croad standard seeks in males and might continue for the females, because I am not fond of any excess of fluff on thighs or sides of abdomen.   I penalise such excess of feathering because I know it must mean so many eggs off during the year.   Eggs come from the excess of food that is left after all repairs have been attended to by way of feather, bone, flesh and so on.   Hence I cut excesses in all these points in order to get nearer to eggs.   Nature puts the latter last and we put them first, so we must help the bird by cutting out excesses. A short-legged bird, too, is soon put out of tone in the winter when running in wet grass, and is handicapped in growing stages.

Passing on to plumage, the Club ideal is soft feather, neither loose nor tight, and in females cushion full but not obtrusive. Just as I do not like handfuls of feathers on thighs, nor do I like " Cochiny " saddles or feathering.   Starting at the base of the neck I like to go down the back to the tail brushing the feathers the wrong way.   In the super-layer the feathers spring back readily, and I like this to happen as I approach the tail.   When the cushion is " rubbed the wrong way " the feathers should not stand upright and be loose as in the Cochin.   Even the Club ideal seems to desire such feathering, but does not define it in any exact terms, but I do not like a cushion that is too full unless it be " on the tight side," i.e., springy.

Another reason why egg-production is not aided by an excess of fluff and feather . . . Does not the fluffy Cochin take longer to moult than the tight-feathered laying type of Leghorn, and does not the former lose many eggs in the season thereby ?   I do not mind an abdomen that is well covered with fluff, but the latter should be silky and pliable to the touch and not coarse.   If there is an excess I always put the fine-textured kind in front of the coarse, and give marks for the former on account of the texture.

Croad type asks for a broad, deep breast, with long, rounded breast-bone.   In a table bird that would be excellent, but my trend is for a shorter straight bone in a laying type to give abdominal capacity, and for a like reason I do not favour too deep a breast. The more we get to the ball-shaped or circular type, as in Show Orpingtons, the more we crowd in the internal organs and lose egg-laying efficiency.   Thus I favour greater development towards the rear and not so much in front, i.e., no extreme.   As will be seen I prefer lengthy types of breeds for laying power as giving more play to internal organs.   With the Croad's medium length of back remaining (and I do not like any shortness here) and its great width of back we can even things up by a roomy abdomen that posteriorly is deep and wide for general capacity. I take my measurements by the skeleton and not by the feathers—a Rhode Island Red is often longer in appearance than it is when handled because the tail is carried on, but as all utility birds are handled, allowances are not made for feather. In judging length of back I take from the ovary (centre of back across the tops of thighs) to the root of the tail.

Carriage of tail may be criticised. I have seen winning birds with tails that have removed all trace of gracefulness—everything sacrificed for the " fan " shape and height. I should prefer a tail not carried too high and furnishings not in excess and not too " feathery " for the reasons already stated.

The Club description of the Croad can actually be that for a layer, viz., " general appearance that of an active, intelligent bird." We all know that some judges go for the very large Croad, and I do favour the smaller pattern that is active and intelligent and graceful. When I put a bird back into the show pen and utter " some timber " as he falls on to his coarse pins, I know he is not active. Therefore I like a shank that is fine and round in bone without surplus flesh at the back. The scales I like many and small and tight. When we examine the shanks of fancy Orpingtons of the coarse kind we notice the larger loose scales which can easily be counted. The scales run in " two's " down the shank instead of " threes " ; my perfect leg is one where there are three scales in the row as against two large ones. Coarseness in leg and bone, and feather and flesh go hand in hand in almost every instance, all calling for repairs ere egg-production gets a look in. I like a rather heavy hen for breeding but a medium bird for laying ; this I can still get because I can have medium weights for pullets for egg-laying, knowing that in their second year, when I wish to breed from them, they will have become heavier. But weight does not mean coarseness, which loses us not only eggs, but also activity. The best layer is always the bird that is busiest and ever on the move. A coarse, heavy bird is more inclined to be squatting on the perch than busy in the litter, and to take a dislike to trap-nests and nest-boxes in general to which she must fly. Again, the blood so readily gets out of order, and blood controls egg-production in more ways than one, while a coarse, heavy hen in getting out of tone falls into moult early and takes longer to get her new plumage, while she gives more trouble from broodiness. She gets fat too, and that means fewer eggs, odd-shaped eggs, and often those soft or thin-shelled eggs which start the egg-eating vice. And she gives trouble from infertility and weak germs and weakly offspring.

Often one is tempted to regard the utility male on a different footing to the female, but that is unsound in my opinion. You eventually mate the cockerel to the hens, so that your progeny must suffer if the male is coarse and the females fine in bone and texture. Exteriors come mainly from the cockerel, and he should never be coarse in head points for instance, or so big and clumsy as to be far from active and intelligent. Size has always been a fancier's aim, but I think it has been carried too far, from my utility view-point.

I have been critical, but I could have taken other Club Standards that would have needed re-writing ; in this case I have not had occasion to criticise many items. I am sure they will be taken as intended—to be helpful and not just critical. " My ideal " in a laying hen when in full flush of production is five-finger capacity for abdomen, three fingers between pelvics, two fingers between root of tail-socket and pelvis, five inches or more across back, five fingers between legs, thin, straight pelvics, fine-textured flesh, nice head points, very large, moist and fine-textured vent, fine textured bone, horn and feather, and medium size. To this I add breed characters, not in the nature of type which I have covered above, but faults in family characters as, for instance, feathered legs in Wyandottes, white in lobe where it is a fault, and so on. And I score my " utility " birds to 200 points, i.e., 70 for capacity, 70 for capability, 10 for ideal size, 10 health, 10 show condition, and 30 breed characters. To the fancier I say, as I always did, " Study egg-laying type " ; and to the utilitarian, " Study breed characters." Out of the total of 140 (70 for capacity and 70 for capability) birds scoring 100 or over marks are graded in as valuable providing they obtain 50 marks minimum for capability, i.e., general texture.

A few words may be useful on feeding Croads. When I have advised people to keep Croads many have written to say that the eggs were small and that they did not do well with the breed. It is useless to blame the latter in such cases, as the fault lies with the feeding and management. A breed like the " table " Croad inherits the natural tendency to fatten and must therefore be fed on sound lines if the best egg-results are to be obtained. If the birds are low grade for laying and coarse in flesh the more careful still must be the feeding.

Present-day methods of management in winter for egg-production are on intensive lines, i.e., the birds (without regard to breed) are confined to scratching sheds on wet days. With that practice I agree, as it is the only one that gives the fullest egg-baskets. But Croads should be kept busy and " doing something " when they are so shut up and not be allowed to squat on the perches or stand about. I find the best plan is to make proper use of the grain feeds, and I prefer to allow $1\frac{1}{2}$ ozs. of grain per bird for breakfast buried deep in the litter, and $\frac{1}{2}$ oz. of grain per bird at 10.30 a.m., also in the litter. And in addition to the ordinary grain mixture I have a special small-grain mixture to advocate. I do not care much of what it consists so long as it is on the lines of, say, a dry chick mixture. The idea is to give the birds the small grains that will take some finding. Supposing I suggest kibbled wheat, wheat tailings, canaryseed, buckwheat and dari. I keep a sack of this in the food store, and as I dole out the grain rations I mix a handful or so of the " scratch feed " mixture with same for each pen of birds.

At midday raw greenery should be provided. I like to nail a piece of half-inch-mesh netting to the side of the house (interiorly), the netting being, say, eight or nine inches deep and fourteen inches long. I nail it tightly on the two sides and bottom with staples first and then battens, and leave the top open like a purse. Then I push the green food in and the birds can pick out small pieces. I like two of these in each pen, just as I like to put the drinking water here, the grit there, and shell somewhere else, and so on, to spread the birds out and make them " do something " when shut in. Also I like drinking vessels on low shelves to encourage activity.

The feeding of Croads should be on the " filling-without-fattening " style, and green food should be given in plenty as it is bulky and yet is mainly water—fills and does not fatten. As a wet mash I like a mixture of 2 lbs. bran, 4 lbs. middlings, 6 lbs. Sussex ground oats (fine, non-husky kind), 2 or 3 lbs. boiled, finely-minced green food or roots (minced before boiling and then strained), $1\frac{1}{2}$ lbs. fish-meal, this being dried off to a nice palatable, crumbly whole, neither stodgy nor sloppy. The mash should bind well and yet break readily into flakes. Layers must enjoy all they are given. Dry off mash with more middlings.

Croads do well on dry mash, but I prefer not to have hoppers too low, and not to let stock get fat, or troubles start from heart affections, apoplexy, etc. In fact, dry mash feeding needs to be in the capable hands of a poultry-man who is observant of condition or tone, and the mixture of meals must again be filling but not fattening. I prefer to allow each bird 2½ ozs. of wet mash, weighed when prepared, and containing boiled green food as advised ; greenery oils the internals and is invaluable where birds are apt to get on the fat side.

Again, with Croads and like natural-fattening breeds I sacrifice eggs as the breeding season comes on by allowing the breeders out on range earlier and not keeping them intensively, my idea being to secure condition and melt away any fat. Handling of the birds for condition is essential, and should be a guide to management. I place my thumb and finger well into abdomen posteriorly and withdraw under gentle pressure ; if the flesh feels stodgy it is coarse, if it hangs on, medium, but if slippery, pliable or " rubbable " then fine in texture. As I get to the skin and pass the internal organs I hold on and can then judge amount of fat under the skin. A little fat in the winter is good as a reserve-store, but in some of my post-mortem birds I often find a layer of " bladder-of-lard " fat more than an inch in thickness. My aim in the above method is to get fertiles and strong germs. Grade out fat birds from breeding pens.

One other point that may be helpful, and I refer to rearing. The early chick is the payer every time in Croads. Those who are on damp soils should get over their difficulties by hatching out a month earlier than is customary ; such birds have age on their side, and reach maturity nicely. Again, the very early pullet starting to lay early, resting for a short time from, say, November, for a partial moult, and coming into lay again in December, is a valuable breeder in all breeds that tend to fatten as age increases. After all, maximum fertility, hatching and rearing results are imperative for all who keep poultry extensively.

Standard of Excellence.

(Revised 1913).

———

MALE.

SHAPE AND CARRIAGE.

An adult bird should be neither high nor low on leg. **Legs** should be sufficiently long to give graceful carriage to the body, which should be well balanced. **"Thigh"** (*tibia-fibula*) rather short, but long enough to let hock stand clear of fluff. **Shanks** medium length, standing well apart. **Breast** broad, deep and full (fuller in old bird), with long breast bone, keel slightly rounded. **Back** medium length, broad and flat across shoulders, saddle well-filling the angle between the back and tail as seen in profile. If the male has sufficient neck and saddle-hackle the back should appear shorter than in female. **Tail** fan-shaped, well spread to right and left, and carried rather high. It should be level with head, when the bird is standing in a position of attention. **Wings** carried high or low, high preferred. **Neck** medium length. **Head** carried well back.

General appearance that of an active, intelligent bird.

SIZE AND BONE.

Size.—In fowls of such remarkable merit for table purposes, size consistent with type must be a great consideration, and an adult cock should weigh not less than 9 lbs. **Bone.**—Medium, or rather fine, in due proportion to size, but subordinate to the amount of meat carried. The cock should have a higher proportion of bone than the hen. **Flesh.**—White. **Skin.**—Thin and white.

PLUMAGE.—Colour and Sheen.

Feather should be rather soft, neither loose nor tight. **Under-colour.**—Dark grey, darker in female. **Surface-colour.**—Dense black with beetle-green gloss upon it, and free from purple or blue tinge, inclining rather towards yellow. White in foot feather is characteristic of the breed and is not a defect. **Neck-hackle** full. **Saddle** rather abundantly furnished with rich hackles. **Wing Coverts** very brilliant. **Tail,** full with plenty of glossy side-hangers, and two sickle feathers on each side projecting some six inches or more beyond the rest. **Thighs** well covered with rather soft feathers.

HEAD AND FEET.

Head small for size of bird, full over the eye. **Face** free of feathers—red. **Comb** single, medium or rather small in size, straight upright, free from side sprigs, thick and firm at the base, becoming rather thin ; fine and smooth in texture, evenly serrated with five or six spikes (five preferred) standing well off from back of head, bright red. **Wattles** brilliant red, fine in quality, and rather small. **Ear-lobes** red, well-developed, pendant, fine in texture. **Eye** brown, the darker the better, but not black—the ideal is the colour of shell of a ripe hazel nut (vandyke brown comes near it)—large and intelligent. **Beak** light to dark horn colour, preferably light at the tip, and streaked with grey. **Shanks** bluish-black (bluish in adult birds) showing pink between the scales, especially on back and inner side of shank. In male bird intense red should show through the skin along outer side at base of shank feathers ; this is highly characteristic. Scales on shanks and toes nearly black in young birds. Shanks feathered down the outer sides (neither too heavily nor scantily), the outer toe feathered. **Toes** four in number ; long, straight, slender ; the web and bottom of foot pinkish white —deeper the pink the better. Black spots on soles of feet a serious fault. **Toe Nails** white.

FEMALE.

Shape.—Hock need not show in adult hen, as hen carries more fluff than cock.

Size.—Not less than 7 lb. when fully grown.

Plumage.—Cushion fairly full, but not obtrusive.

Tail in adult hen may have two feathers slightly curved and projecting about one inch beyond the rest. In other respects the hen resembles her mate, as detailed above.

Disqualifications.—Yellow legs or feet; yellow in face at base of beak or in edge of eye-lids; five toes; other than single comb; permanent white in ear-lobes; grey (light slate colour) in webbing of flights; more than a mere trace of dark colour or black in toe nails; black or partially black soles of feet, as distinct from black spots.

POINTS.

Type and Condition.—Shape 15; Condition and alertness 10 25

Substance.—Size 15; Bone 10 25

Feather.—Colour and Sheen 15; Furnishings of Tail, etc., 10 25

Head and Feet.—Feet and Nails 10; Footings 5; Eye 5; Comb 5 25

——

100

Approximate Values.—" Exceptional " 25; " Very Good " 20; " Good enough for Stock Bird " 15; " Fair " 10; " Unsatisfactory " 5 to 0.

Note that it is left to the discretion of the judge to penalise **any** bad fault to the extent of 25 points.

Hints to Breeders and Judges on Deviations from the Standard.

(*a*) **Not Objectionable** in stock birds, and not seriously against the bird in show pen. (Judges must use their discretion.)

1.—Purple or blue barring in a very few feathers only, where the others are of good colour.

2.—Dark red (like colour of Rhode Island Red) in a few feathers in neck hackle or on shoulders in male only (believed to arise from excess of green, and cock showing it is suitable to mate with dull hens).

3.—White (not grey) in flights and secondaries ; white tips on head on adult hen only ; white tips or edging on breast in chicken feather only.

4.—The presence of a moderate amount of feathering on middle toe.

(*b*) **Highly Objectionable,** and to be firmly discouraged, but not necessarily indicating impurity.

1.—An appreciable amount of purple or blue barring.

2.—Decided purple or blue tinge.

3.—Light eye (but make some allowance for age).

4.—Yellow iris of eye.

5.—Wry tail and squirrel tail.

6 —Marked scarcity or absence of leg and foot feather.

Edward Cocker

CLUB SHOW JUDGE.

Breeder and Exhibitor of

CROAD LANGSHANS

A limited number of Sittings for SALE.
Headed by some of the most typical Male
birds in the fancy.

101, TOWNGATE,
LEYLAND,
LANCASHIRE.

HERBERT P. MULLENS

The Vicarage, Abbot's Bickington,

BRANDIS CORNER, N. DEVON.

BREEDER & EXHIBITOR
OF
HIGH-CLASS
Croad Langshans.

The Dual-Purpose Strain.

□ □ □

EGG AVERAGE:
Over **200** per annum.

□ □ □

Combined with Perfect Type.

BIRDS AND EGGS USUALLY FOR SALE.

R. O. RIDLEY

Docking Hall,
KING'S LYNN,
NORFOLK.

BREEDER OF

Croad Langshans

for over 25 years.

BIRDS AND EGGS DIRECT FROM LATE MISS CROAD.

My strain are layers of dark brown eggs, and a Pullet shown at the N.U.P.C. Show scored 168½ marks out of a possible 200.

CROAD LANGSHANS

Type with Utility qualities.

Special attention to breeding for
SIZE and COLOUR of EGG.

**EGGS FOR HATCHING.   A FEW STOCK BIRDS.
NO DAY-OLD CHICKS.**

ALEX. SMITH

. . . *Schoolhouse,* . . .

Telegrams: Schoolhouse, Forgue.	FORGUE, HUNTLY, ABERDEENSHIRE.	Station: Huntly, G.N.S. Rly

King Alfred: and Other Poems.

By SARDIUS HANCOCK.

Crown 8vo. 5s. net.

An historical drama in Tennysonian blank verse which displays an unusual measure of poetic force and literary charm. A real poetic impulse drives the author's pen. The lyrics are graceful and musical and there is a present-day note which will appeal to those who are not usually readers of poetry.

"Mr. Hancock is quite an accomplished writer of verse. . . . In lyric, ballad or sonnet are heard continually echoes of the masters. . . ."—*The Times.*

"Mr. Hancock possesses striking poetic talents. 'King Alfred' is likely to be considered by competent dramatic critics as worthy of production on the stage."—*Malvern Gazette.*

"The volume is worthy of a place in every home."
—*Worcestershire Echo.*

PHILIP ALLAN & CO.,
QUALITY COURT, CHANCERY LANE,
LONDON, W.C. 2.